Essentials
of
Lean *Six Sigma*

Essentials
of
Lean *Six Sigma*

Salman Taghizadegan

AMSTERDAM • BOSTON • HEIDELBERG • LONDON
NEW YORK • OXFORD • PARIS • SAN DIEGO
SAN FRANCISCO • SINGAPORE • SYDNEY • TOKYO
Academic Press is an imprint of Elsevier

Butterworth–Heinemann is an imprint of Elsevier
30 Corporate Drive, Suite 400, Burlington, MA 01803, USA
Linacre House, Jordan Hill, Oxford OX2 8DP, UK

 Recognizing the importance of preserving what has been written, Elsevier prints its books on acid-free paper whenever possible.

Library of Congress Cataloging-in-Publication Data
Application submitted

British Library Cataloguing-in-Publication Data
A catalogue record for this book is available from the British Library.

ISBN 13: 978-0-12-370502-0
ISBN 10: 0-12-370502-9

For information on all Butterworth–Heinemann publications
visit our Web site at www.books.elsevier.com

Printed in the United States of America
06 07 08 09 10   10 9 8 7 6 5 4 3 2 1

To my loving wife, Leila and our daughters, Sara and Setareh.
To my father and my late mother who asked so little and gave so much.

Contents

PART I
Statistical Theory and Concepts

Chapter 1
Introduction to Essentials of Lean *Six Sigma* (6σ) Strategies

Chapter 2
Statistical Theory of Lean *Six Sigma* (6σ) Strategies

Chapter 8

Road Map to Lean *Six Sigma* Continuous Improvement Engineering Strategies

PART III
Case Studies

Chapter 9

Six Sigma Green and Black Belt Level Case Studies

Chapter 10
Six Sigma Master Black Belt Level Case Study

Preface

ABOUT THIS BOOK

Before the 1970s, the industry standards were based on $\pm 3\sigma$ and percent (%) defect. Now, as population grows and industrial volume due to global economy becomes mass production, the $\pm 3\sigma$ and percent defect evaluations are no longer valid. The $\pm 6\sigma$ and defect per million are today's standard for ultimate customer satisfaction and maximum profitability. Knowing that customer satisfaction is the number one priority on any organization's list, the success of any company depends on quality and competitive product pricing. Today the globalized market allows no space for error. Thus, *Six Sigma* is necessary for all organizations. The theory of *Six Sigma* demonstrates the bottom line and customer satisfaction improvement.

Unlike other programs that concentrate on quality only, *Six Sigma* focuses on customer satisfaction and the bottom line. This also means the highest quality: As defects drop to 3.4 per million, quality improves dramatically.

This book explains the Lean *Six Sigma* concepts, the essential theory and analysis from the engineering point of view in three different parts. Part I: Statistical Theory and Concepts; Part II: *Six Sigma* Engineering and Implementation; and Part III: Case Studies. Throughout this book numerous examples have been cited, particularly in the plastics industry of injection molding. All other manufacturers may also benefit to a great extent. Consequently, any other organization may engineer their *Six Sigma* program using this book, as well. A brief description of each chapter follows:

Part I. Statistical Theory and Concepts
Chapter 1. Reviews Lean *Six Sigma* concepts and background (history).
Chapter 2. Demonstrates normal distribution, process capability estimation of 1 sigma through *Six Sigma*.
Chapter 3. Explains essentials of mathematical concepts in Lean *Six Sigma* engineering strategies, as well as a review of standard normal distribution and normality tests.

Part II. *Six Sigma* Engineering and Implementation
Chapter 4. The essentials of *Six Sigma* continuous improvement principles and training models.
Chapter 5. The essentials of design for *Six Sigma* principles, tools, and techniques.
Chapter 6. The essentials of design for Lean *Six Sigma* and training models.
Chapter 7. The roles and responsibilities to *Six Sigma* philosophy and *Six Sigma* infrastructure.
Chapter 8. The road map to Lean *Six Sigma* continuous improvement engineering strategies.

Part III. Case Studies
Chapter 9. Case studies with complete *Six Sigma* applications in injection molding plastics manufacturing for Green and Black Belts.
Chapter 10. An expanded version of the case study published by the author for the Black Belt level in reduction or minimization of variation in an injection molding plastic industry.

Salman Taghizadegan

Acknowledgments

The author wishes to thank Keith Boyle (Quality & Productivity Resources) for his inspiration, assistance, and contributions throughout this book. He also provided my Black Belt training at the University of California at San Diego. Much appreciation goes to my family. To my wife, Leila, and my daughter, Sara, thank you for your long-lasting patience and continuous support. I would like to acknowledge Hunter Industries for supporting and giving me the opportunity to implement Lean *Six Sigma* concepts in practice. I thank Diane Clark for her support of the publication of this book (and for putting in long hours and making this book easier to read). Finally, my thanks to Joel Stein at Elsevier Science for his support and patience throughout the publishing process and other members of the Elsevier Science team for their support and assistance in making this work a reality. My thanks to Shelley Burke for her assistance throughout the publishing process. Furthermore, thanks goes to Carl M. Soares for managing the Publication process.

About the Author

Dr. Salman Taghizadegan has substantial experience in chemical and plastics processing, design, control, and analysis. He received his B.S. in chemistry from Western Illinois University, his B.S. in chemical engineering from the University of Arkansas, his M.S. in chemical engineering from the Texas A&M University, and his Ph.D. in chemical engineering with emphasis in plastics from the University of Louisville.

He has over 20 years of academic and full-time industrial experience in plastics and chemical processing, design, and control engineering, primarily in injection molding industries. He has authored numerous technical publications and has spent most of his professional career as an adjunct professor in engineering, as a highly technical specialist in the plastics industry, as a leader in quality and process improvement, and as the manager of waste reduction in the manufacturing environment. Dr. Taghizadegan is certified Six-Sigma Black Belt and Master Black Belt through the University of California at San Diego and the university of San Diego. He is a member of the Society of Plastics Engineers.

Chapter 1

Introduction to Essentials of Lean *Six Sigma* (6σ) Strategies

Lean *Six Sigma*: *Six Sigma* Quality with Lean Speed

1.1 LEAN *SIX SIGMA* (6σ) CONCEPT REVIEW

1.1.1 THE PHILOSOPHY

In any organization customer satisfaction is the number one priority. Customer satisfaction also means profitability. The success of any company depends on the ability to ensure the highest quality at the lowest cost. In the 1980s when most companies believed that producing quality products was too costly, Motorola believed the opposite: "the better, the cheaper." It realized that by producing a higher-quality product, the cost of producing goes down. Motorola knew that greater customer satisfaction generates higher profitability.

Today the competitive market leaves no space for error. It is now necessary to implement the concepts of Lean *Six Sigma*. Lean *Six Sigma* is a business strategy in which the focus is to improve the bottom line and increase customer satisfaction.

Six Sigma philosophies are related to statistical process control, stochastic control (relating to probability), and engineering process control. In addition, it requires process and data analysis, optimization methods, lean manufacturing, design of experiment, analysis of variance, statistical methods, mistake-proofing, on-time and or on-schedule shipping, waste reduction, and consistency assurance. It is a process capability that continuously improves the quality of the product and maximizes productivity. In simpler terms, Lean *Six Sigma* is the following:

1. It is a data-driven approach and methodology to analyze the root causes of manufacturing and business problems/processes by eliminating defects (driving toward six standard deviations between the mean and the nearest specification limit), and dramatically improving the product.
2. It improves the employee's knowledge of business management to distinguish the business from the bottom line, customer satisfaction, and on-time delivery. Thus, *Six Sigma* is not just process-improvement techniques but a management strategy to manage the projects to financial goals.
3. It combines robust design engineering philosophy and techniques with low risks (Lean *Six Sigma* tools: measure, analyze, develop, and verify).

It would be very difficult to achieve this goal without teamwork and proper training of the entire organization to a higher level of competency. During the 1980s *Six Sigma* grew into a distinct manufacturing discipline. It now encompasses a wide range of disciplines, including transportation, administration, manufacturing, medical, and a variety of other operating organizations and processes (by definition a process is any operation that has an input and produces an output).

1.1.2 LEAN/KAIZEN *SIX SIGMA* ENGINEERING

Lean speed is a technique as well as a continuous effort that is used to accelerate and minimize the cost of any process by eliminating the waste in either manufacturing or service. Basically, Lean philosophy identifies and removes inefficiencies like the nonvalue-added (waste) cost or unneeded wait time within the process caused by defects, excess production, and other processes to expand any organization. For example, in most cases 95% of the lead time (from the beginning to the end of a process) is the wait time. Further, 80% of process delays are caused by a 20% time trap (activities in the workstation). By improving 20% time trap, it can eliminate 80% of process delays. Hence, Lean is associated with speed, efficiency, and acceleration of the process. Therefore, by integrating elements of Lean enterprise methodology with *Six Sigma*, which lacks tools that control and reduce lead time, the feedback will be faster than planned.

The combination of these two powerful tools, Lean manufacturing and *Six Sigma* strategy, will result in process variation reduction and dramatic bottom-line (language of CEO) improvement. Since all companies are in the business of achieving faster return on investments, particularly for their shareholders, using Lean principles in *Six Sigma* is extremely important. For the company architecting *Six Sigma* philosophy in its infrastructure, Lean manufacturing speed can accelerate the implementation and benefits of the manufacturing process.

Here are some of the basic Lean manufacturing techniques and principles that are used in Lean *Six Sigma*:

1. 5S
    - Sort (keep things that are essential), Shine (keep everything clean), Straighten (make everything visible and accessible), Standardize (implement the first 3S and maintain them), and Sustain.
    - The first 3S are actions, and the last two are sustaining and progressive.
2. Value-stream mapping
    - A method of mapping a product's production path from manufacturing facility to customer's door.
    - A visual tool for identifying all steps of operations in the manufacturing process with cost-effective results.
3. Kaizen event
    - Continuous improvement.
4. Mistake-proofing
    - Process analysis and implementation of robust engineering to build quality into an assembly or manufacturing process with cost-effective results.
5. Cycle time reduction
6. Inventory reduction
7. Setup time reduction
8. Waste identification and elimination

In other words, the Lean speed is merged with or is embedded within the *Six Sigma* principles. The integration of these two concepts will both deliver faster results and achieve the best competitive position by concentrating on the use of tools that have the highest impact on the already established performance levels. Another example is the design of experiment that may require about 16 runs to determine optimum factors and reduce variation. Minimizing the lead time by 80% will allow the experiment to be completed five times faster using fractional factorial design. Basically Lean contributes to *Six Sigma* in the following manner:

1. Eliminates all the waste time that slows the project.
2. Maintains customer satisfaction with speed in delivery.
3. Gets the project done under the deadline and possibly under budget.
4. Continuously improves the profitability (e.g., in a shorter period of time than planned).

1.2 *SIX SIGMA* BACKGROUND

Motorola engineering scientist William Smith, known as the father of *Six Sigma*, developed the concept in the 1980s. For many years, he and other pioneering engineers and scientists worked on this or similar concepts to reduce variation, improve quality, and maximize productivity, including Walter A. Shewhart, W. Edwards Deming (see Appendix for Demingr's 14 points for management), Philip R. Crosby, Shiego Shingo, Taiichi Ohno, and Joseph Juran. Each one studied quality from a different angle.

The methodology of *Six Sigma* uses the statistical theory and thus assumes that every process factor can be characterized by a statistical distribution curve. The objective is to free all the defects from every process, product, and transaction. It is a process that provides tools to achieve nearly error-free products and services with maximum profitability. In the 1960s and 1970s, statistical process control limits were based on plus or minus three sigma (±3 standard deviation) from the mean. However, in this concept the process limits are plus or minus *Six Sigma* from the mean.

Just like three sigma, *Six Sigma* is applicable to batch-to-batch process, discrete, and continuous applications. The goal is to produce less than four defects per one million operations. *Six Sigma* will enable a company to capture substantial market share in the competitive global markets. Global competitiveness almost becomes impossible without *Six Sigma*. Every company would benefit by adopting *Six Sigma* concepts and philosophy. Profitability improves tremendously if it is applied to all workforces in every department of the corporation.

1.3 *SIX SIGMA* SUCCESSES

An example of a *Six Sigma* successe is Motorola Corporation, which increased net income from $2.3 billion in 1978 to $8.3 billion in 1988, using the *Six Sigma* program. As a result, Motorola received the Malcolm Baldrige National Quality Award by President Reagan in 1988. The award is presented to the industries that become quality role models for others. GE also implemented *Six Sigma* in the mid-1990s in a five-year program and boosted its profits by a substantial amount. By the year 2002 GE had achieved $4 billion in savings per year. Other companies that benefit from *Six Sigma* are Allied Signal, Inc.; Polaroid Corporation; Asea Brown Boveri Power Transformer Company; and DuPont.

At three sigma the cost of quality is 25 to 40% of sales revenue. At *Six Sigma* it reduces cost of quality to less than 1% of sales revenue. In fact, Lean *Six Sigma* is the epitome of quality and should be adopted by all manufacturing companies to remain in business. Therefore, one must change measurement of quality in parts per hundred (percentages) to parts per million. This has changed the makeup and culture of industries that adopted Lean *Six Sigma*.

Sigma Variation

Mathematically variation and reproducibility are inversely related to each other—for example, as variation increases, producibility decreases due to increase

Table 1.1

Comparisons of 3.8 Sigma and *Six Sigma* Defect Examples

3.8 Sigma (99% Good)	*Six Sigma* (99.99966% 6σ)
• 200,000 wrong drug prescriptions per year	• 680 wrong prescriptions per year
• 5,000 incorrect surgical operations per week	• 88 incorrect operations per week
• More than 15,000 newborn babies accidentally dropped per year	• 5 newborn babies dropped per year
• 2 short or long landings at major airports per day	• Less than 1 short or long landing every 8 years
• 20,000 articles of mail lost per hour	• 7 articles lost per hour

Table 1.2

Comparisons of Old (Traditional) and New (Lean *Six Sigma*) Methods

Problem	Old methods	New methods
Design	Product performance	Product producibility
Analysis	Experience based	Data based
Issue	Fixing problems	Preventing problems
Manufacturing/ Molding	Trial & error process	Robust design process
Inventory level	High production quantity	Low production quantity as needed
People	Cost to company	Asset to company
Management	Cost & time	Quality & time
Employee goal	Company	Customer
Product engineering	Little input from customer	High input from customer
Quality focus	Product	Process
Dominant process factors—selection	Apply one factor at a time	Apply design of experiment
Process improvement	Robotic technique	Optimization technique
Proving	Experience based	Statistically based
Company outlook	Short-term plan	Long-term plan
Customer satisfaction	Production at statistical acceptance quality level	Fewer defects, when and what quantity customer wants
External relationship	Price relationship	Long-term relationship
Layout	Functional	Cell type
Production schedules	Forecast	Customer order
Manufacturing cost	Continuously rising	Stable and decreasing

of nonconformance (in the technical sense called a rejection or defect) probability. Additional workforce, cost, scrap, and cycle time reduce the sigma level where such variation comes from design, process, and material of the finished products. Consequently, sigma variation reduces customer satisfaction and has negative impact in the profitability, which is one of the main focus areas of *Six Sigma*.

It is too difficult to convert any operation from three sigma (3.0σ) to *Six Sigma* (6.0σ) in one step. It will require several steps of improvements from 3.0σ to 4.0σ, 4.5σ, 5.0σ, 5.5σ, and finally 6σ (Tables 2.2 through 2.4 illustrate how as sigma increases, product quality and profitability also increase). This also means that cycle time is reduced, quality checks are minimized, operating cost goes down, variable costs shrink, and customer satisfaction goes up. At *Six Sigma*, all products conform to a worldwide standard and are nearly defect free. In other words, *Six Sigma* determines the capability of the process to accomplish a defect-free work environment. So sigma range dictates how often defects are likely to happen in the system. *Six Sigma* is not twice as good as three sigma but almost 20,000 times better.

Some examples of *Six Sigma* quality for long-term processes are shown in Table 1.1.

Highlights of some of the *Six Sigma* cultural changes are listed in Table 1.2.

Chapter 2

Statistical Theory of Lean *Six Sigma* (6σ) Strategies

2.1 NORMAL DISTRIBUTION CURVE

The concept of the normal distribution curve is the most important continuous distribution in statistics. The normal distribution curve plays a key role in statistical methodology and applications. For instance, suppose for each of six days samples of 11 parts were collected and measured for a critical dimension concerning a shrinkage issue. The number of parts with dimensions is listed in Table 2.1.

Figure 2.1 illustrates the graphical representation of frequency distribution for the data in Table 2.1, having an upper specification limit (USL) 0.629, mean 0.625, and lower specification limit (LSL) 0.621 (tolerance = ±0.004). This means that any data above 0.629 and below 0.621 are assumed defects (out of specification). Figure 2.1 indicates that data (population) are symmetrically distributed. By locating bullets on the middle top of each column (as shown in Figure 2.2) and connecting them, we set a bell-shaped curve, as in Figure 2.3, which is also called a normal distribution curve (Figure 2.4). (This is discussed in detail in Section 3.2.) The area under the distribution curve is the probability of variations from the mean of any process.

2.2 *SIX SIGMA* PROCESS CAPABILITY CONCEPTS

Any manufactured part (or finished product) is considered scrap if it does not meet the required measurement (e.g., dimensional, physical, mechanical, or chemical properties). In other words, anything that results in customer dissatisfaction is called a defect. What this tells us is that the value is outside the

Table 2.1

Shrinkage Study Samples

Row	Number of parts	Dimensions
1	1	0.620
2	3	**0.621 = LSL**
3	5	0.622
4	8	0.623
5	10	0.624
6	12	**0.625 = Mean**
7	10	0.626
8	8	0.627
9	5	0.628
10	3	**0.629 = USL**
11	1	0.630

LSL = 0.621, Mean = 0.625, USL = 0.629

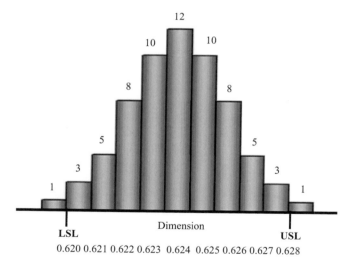

Figure 2.1   Histogram of sample dimensions.

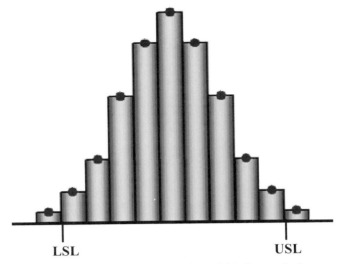

Figure 2.2   Histogram of sample dimensions with bullets on the frequency.

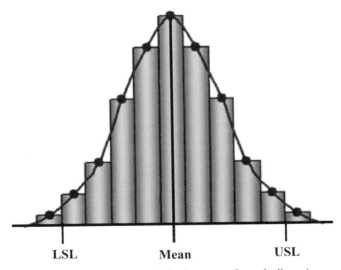

Figure 2.3   Histogram and distribution curve of sample dimensions.

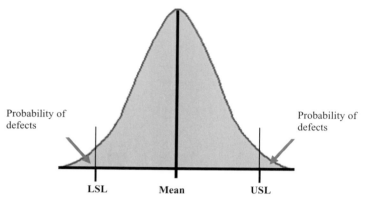

Figure 2.4   Distribution curve of sample dimensions.

customer specification limits—that is, the lower specification limit (LSL) and upper specification limit (USL). The LSL and USL are determined by customer requirements. The center in between LSL and USL is known as the mean or target value.

Once a production process is in control (using statistical process control methods), the question is, can the process maintain its capability for a longer period of time? To determine this, one may investigate the short- and long-term capability of a desired process. The difference between short and long capability is shown in the following sections. (See Chapter 8 for more on process capability.)

2.2.1 *SIX SIGMA* SHORT-TERM CAPABILITY

Six Sigma short-term capability occurs when the process is centered on the target and there is no distribution shift. It also assumes continuous uniform process with no changes. The process capability for the short term is shown in Tables 2.2 and 2.3.

For simplicity, Figure 2.5 illustrates Table 2.2 in x and y coordinates (magnitude).

Table 2.3 illustrates the values of Table 2.2 in percentages.

The theoretical value of 100% cannot be reached in practice because the curve meets the x-axis in infinity. The statistical representation (distribution curves) of Table 2.2 for three sigma capability and *Six Sigma* capability when the process is centered on the target is shown in Figure 2.6.

Figure 2.7 illustrates the distribution curve/ process and design width for *Six Sigma* capability when the process is centered at the target.

Table 2.2

Impact of Process Capability of One Sigma through *Six Sigma* for Short Term When the Process Is Centered on the Target

Sigma capability	Defect free/million	Defect/million (expected nonconformances)
1.0 Sigma	682,690	317,310
2.0 Sigma	954,500	45,500
3.0 Sigma	997,300	2,700 (Traditional quality)
3.5 Sigma	999,535	465
4.0 Sigma	999,937	63
4.5 Sigma	999,993.2	6.8
5.0 Sigma	999,999.4	0.6
6.0 Sigma	999,999.998	0.002 (2 parts per billion)

Table 2.3

Mathematical Comparison of Sigma Capability Concepts for Short Term

Sigma capability	Defect free per million	Defect per million	Quality/ profitability
1.0 Sigma	68.269000%	31.731000%	loss
2.0 Sigma	95.450000%	04.550000%	
3.0 Sigma*	99.730000%	00.270000%	an industry average
3.5 Sigma	99.953500%	00.465000%	
4.0 Sigma	99.993700%	00.006300%	above average
4.5 Sigma	99.999320%	00.000680%	
5.0 Sigma	99.999940%	00.000060%	
6.0 Sigma	99.9999998%	00.0000002%	

*Traditional quality

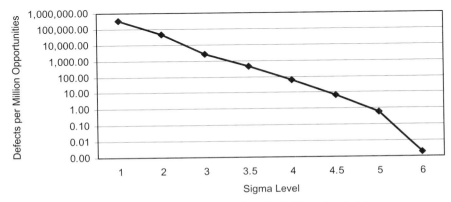

Figure 2.5   Capability of one sigma through *Six Sigma* for short term.

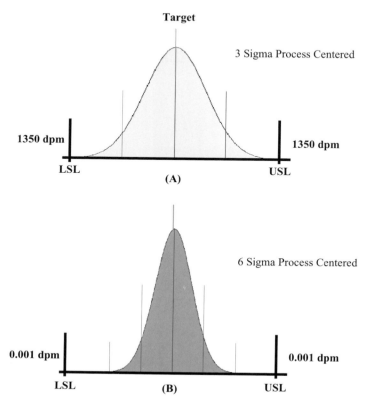

Figure 2.6   **A.** Three Sigma capability. **B.** *Six Sigma* capability. When the process is centered on the target.

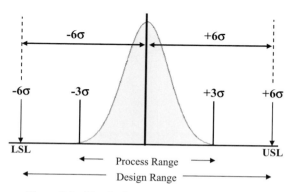

Figure 2.7   Distribution curve centered at the target.

2.2.2 ESTIMATION OF *SIX SIGMA* LONG-TERM CAPABILITY

Six Sigma assumes that the process mean changes from lot to lot, especially within a large quantity of lots. In addition, this change could result in an average of 1.5 sigma distribution shift in either direction for long-term performance. This is due to the fact that operator error and machine wear and tear contribute to the offset of sigma from the target line. This is demonstrated in Table 2.4 for long-term operation. Using this concept one can use an Excel spreadsheet and Equation 2.1 to calculate the sigma capability for various nonconformance probabilities of product yield.

$$\text{Sigma Capability} = \text{NORMSINV (Probability)}$$

$$\text{Sigma Capability} = \text{NORMSINV}\left(1 - \frac{Defects}{10^6}\right) + 1.5 \qquad (2.1)$$

where NORMSINV stands for the inverse of the standard normal cumulative distribution. Basically, NORMSINV uses an iterative method for evaluating the function. By giving a probability value [1 − (Total defect/total opportunity or population)], NORMSINV iterates until the result is correct to within ±3E-7. If NORMSINV does not converge after 100 iterations, the function responds as an error message. For instance, if in a given problem defect counts in a population of 10E-6 parts are 150, then sigma level per million using Excel can be calculated as

Table 2.4

Impact of Process Capability of One Sigma through *Six Sigma* for the Long Term When the Process Is Offset by 1.5 Sigma

Sigma capability	Defect free per million	Defects per million
0.0 Sigma	67,000	933,000
1.0 Sigma	310,000	690,000
1.5 Sigma	500,000	500,000
2.0 Sigma	691,700	308,300
2.5 Sigma	841,350	158,650
3.0 Sigma	933,193	66,807 (Traditional quality)
3.5 Sigma	977,300	22,700
4.0 Sigma	993,780	6,220
4.5 Sigma	998,650	1,350
5.0 Sigma	999,767	233
5.5 Sigma	999,968	32
6.0 Sigma	999,996.60	3.40

$$\text{Sigma capability} = \text{NORMSINV}\left(1 - \frac{150}{10^6}\right) + 1.5 = 5.12 \quad \text{or} \quad \sigma = 5.12$$

The full sigma conversion table defects per million opportunities (DPMO) to sigma to percent yield is given in Appendix IV.

In practice one can also determine DPMO by dividing total defects by the total opportunities and multiplying the results by a million.

$$\text{DPMO} = \frac{Defects}{Opportunities} \times 10^6$$

For example, if one makes 650 parts and 20 parts are defective, then

$$\text{DPMO} = \frac{20}{650} \times 10^6 = 30{,}769$$

and sigma = 3.37, defect % = 3.08, yield = 96.92.

Figure 2.8 illustrates the capability of one sigma through *Six Sigma* for the long term when the process is offset by 1.5 sigma.

Thus, Tables 2.2 and 2.3 show that the short-term process will produce no more than 0.002 defects per million when sigma is equal to 6. In the long term this amount will increase to 3.40 when sigma is equal to 6. Figure 2.9 is the graphical representation of Table 2.4 for three sigma capability and *Six Sigma* capability when the process distribution is shifted by 1.5 sigma from the center or target.

Figure 2.10 shows the process and design width for *Six Sigma* capability when the process is shifted 1.5σ from the target.

In practice the shift could go either to the left (negative side) or to the right (positive side) of the target. It cannot be on both sides at the same time. As long

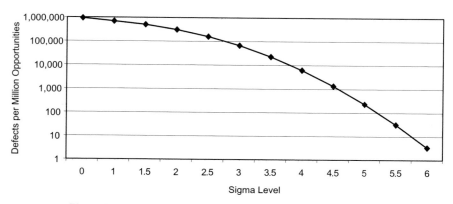

Figure 2.8 Capability of one sigma through *Six Sigma* for long term.

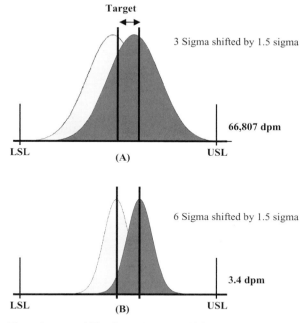

Figure 2.9 **A.** Three sigma capability. **B.** *Six Sigma* capability. When the process shifted 1.5 sigma from target.

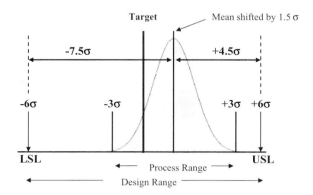

Figure 2.10 Distribution curve shifted by 1.5σ to the right of target.

as the shift remains within the design limits (customer requirements), the process will produce very minimum defects or none. Otherwise, the process will require tuning (or adjustment). Figure 2.11 shows that the process can go in either direction of the target by 1.5σ but only one direction at a time.

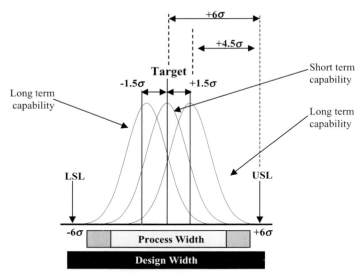

Figure 2.11   Distribution curve shifted by ±1.5σ to the right or left of the target.

Table 2.5 illustrates the values of Table 2.4 in percentages. Percent yield per million can be calculated using the following formula:

$$\text{Yield } \% = \left(\frac{1.0E\text{-}6 - Defects}{1.0E6} \right) \times 100$$

Lean *Six Sigma* (6σ) contains 99.9997% of all values. In fact, as mentioned in the foregoing discussion, when the process sigma value increases from zero to six, the variation of the process around the mean value decreases. With a high value of process sigma, the process approaches very minimum variation and tends to almost zero defects. This is shown in Figure 2.12.

All the defects are the outcome of improper design, process, material, or machine operation. Figure 2.13 illustrates this concept.

The outcome of all of the preceding improving techniques is also illustrated in the control chart in Figure 2.14 (see Section 8.7.1.2). Thus, capability analysis concludes that *Six Sigma* means world class products, customer satisfaction, cheaper, and even quicker than before. It also means variations are not acceptable.

Chapter 3 describes the mathematical concepts of *Six Sigma*, probability distributions, and reliability. It reviews the mathematical point of view of *Six Sigma* explained in this chapter, as well as models and functions. Mathematically no process can be understood or measured and/or controlled without converting it

Table 2.5

Mathematical Comparison of Sigma Capability Concepts for Long Term

Sigma range	Percent defect-free per million	Defect per million	Quality/profitability
1.0 Sigma	31.0000%	69.0000%	loss
1.5 Sigma	50.0000%	50.0000%	
2.0 Sigma	69.1463%	30.8537%	noncompetitive
2.5 Sigma	84.1350%	15.8650%	
3.0 Sigma	93.3193%	06.6807%	average industries
3.5 Sigma	97.7300%	02.2700%	entering above average
4.0 Sigma	99.3790%	00.6210%	above average
4.5 Sigma	99.8700%	00.1400%	
5.0 Sigma	99.9767%	00.0233%	below maximum productivity
5.5 Sigma	99.9968%	00.0032%	
6.0 Sigma	99.9997%	00.0003%	near perfection/near maximum profit

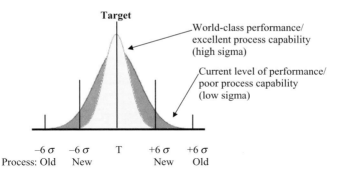

Figure 2.12   As the process sigma value increases from zero to six, the variation of the process around the mean value decreases, which means process capability improves.

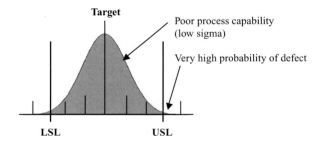

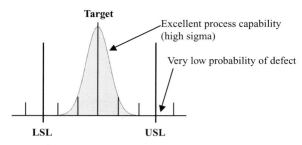

Figure 2.13   Capability and reliability.

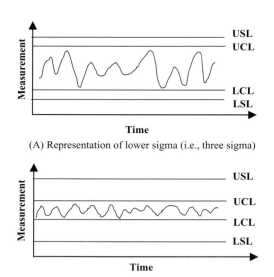

Figure 2.14   Illustration of control chart in lower sigma and higher sigma.

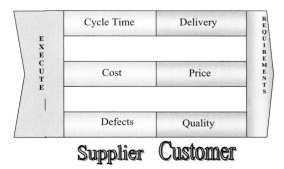

Figure 2.15 Supplier and customer relation.

to the form of real numbers. In general, the processes can be transformed into mathematic expressions.

The customer requirements, such as on-time delivery, lower prices, and higher quality, will be met by executing lower cycle time, cost reduction, and minimum defects. This is shown in Figure 2.15.

As mentioned before, customer satisfaction is a function of quality, cost, and delivery (see Section 6.6).

Chapter 3

Mathematical Concepts of Lean *Six Sigma* (6σ) Engineering Strategies

3.1 PROCESS MODELING—THE HEART OF LEAN *SIX SIGMA*

As mentioned in Chapter 1, variation is inversely related to identical predictability—for example, as variation increases, reproducibility decreases due to increase of nonconformance probability. This results in additional cost, defects, cycle time, lead time, extra production, inspection, Takt-time (or cycle time), and so on, which reduce the sigma capability. Such variation comes from design, process, tooling, process layout, and extra work-in-process (WIP) and material of the finished products. In other words, product performance is a function of design, material, tooling, process condition, and process speed (lead time), as shown in Figure 3.1. In mathematical terms, let the value of the dependent variable response Y change as a function of independent variables (or adjustable variables) $x_1, x_2, x_3, \ldots, x_n$ of n quantitative factors such that

$$\text{Output} = f(\text{Input})$$

or

$$Y = f(x) = f(x_1, x_2, x_3, \ldots, x_{n-2}, x_{n-1}, x_n) \tag{3.1}$$

Where f is called response function and Y is a dependent variable (in *Six Sigma* terms strategic objective measurement values or symptom) and x is an independent variable (or cause). In the case of design, material, tooling, and process condition, Equation 3.1 becomes

$$Y = f(x_1, x_2, x_3, x_4) \tag{3.2}$$

Y is the response or product performance, and x_1, x_2, x_3, and x_4 represent design, material, tooling, and process condition, respectively.

Essentials of Lean *Six Sigma*
Copyright © 2006 by Academic Press, Inc. All rights of reproduction in any form reserved.

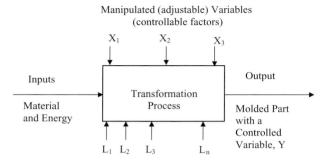

Manipulated (adjustable) Variables
(controllable factors)

Figure 3.1   General model of a transformation process.

However, by looking at design, material, tooling, and process condition, one may find that these factors are also dependent variables. This means they are a function of other variables. Consequently, in this case Equation 3.2 becomes

$$Y = f(y_1, y_2, y_3, y_4) \tag{3.3}$$

Equation 3.3 is expanded as in Equations 3.4 to 3.7.

Product design $= f$ (dimensions, plastic properties, etc.)

or

$$y_1 = f(x_1, x_2, x_3, \ldots, x_{n-2}, x_{n-1}, x_n) \tag{3.4}$$

Material behavior $= f$ (viscosity or melt index, density, shrinkage, impact, mechanical properties, modulus, molecular distribution, etc.)

or

$$y_2 = f(x_1, x_2, x_3, \ldots, x_{n-2}, x_{n-1}, x_n) \tag{3.5}$$

Tooling $= f$ (sprue, runner, gate size, vents, cooling, core, cavity, steel type, etc.)

or

$$y_3 = f(x_1, x_2, x_3, \ldots, x_{n-2}, x_{n-1}, x_n) \tag{3.6}$$

Process $= f$ (pressure, temperature, time, velocity, shear rate, volume or injection stroke, clamp tonnage)

or

$$y_4 = f(x_1, x_2, x_3, \ldots, x_{n-2}, x_{n-1}, x_n) \tag{3.7}$$

As mentioned, the $x(s)$ is also called an independent (cause) and $y(s)$ a dependent (symptom) variable. The polynomial representation of Equations 3.2–3.6 at any point $(x_1, x_2, x_3, \ldots, x_{n-2}, x_{n-1}, x_n)$ in the factor space, or n-dimensional space, can be represented in the regression equation form

$$y_n = b_0 + b_1 x_1 + b_2 x_2 + \cdots + b_{11} x_1^2 + \cdots + b_{12} x_1 x_2 + \cdots + b_{111} x_1^3 + \cdots \text{ etc.}$$

And in the summation notation form

$$y_n = b_0 x_0 + \sum_{i=1}^{n} b_i x_i + \sum_{i=1}^{n} b_{ii} x_i^2 + \sum_{\substack{i=1 \\ i<j}}^{n-1} \sum_{j=2}^{n} b_{ij} x_i x_j + \sum_{\substack{i=1 \\ i<j<l}}^{n-2} \sum_{j=2}^{n-1} \sum_{l=3}^{n} b_{ijl} x_i x_j x_l + \cdots \text{ etc.}$$

$$(3.8)$$

where x_0, a dummy variable, is always equal to unity; $x_1, x_2, x_3, \ldots, x_n$ are the independent variables that affect the y value; b_0, b_i ($i = 1, 2, \ldots, n$), b_{ij} ($i = 1, 2, \ldots, n; j = 1, \ldots, n$), b_{ijl} ($i = 1, 2, \ldots, n; j = 1, \ldots, n; l = 1, \ldots, n$), $\ldots$ and so on are unknown independent coefficients. In the case of three factors x_1, x_2, and x_3, the corresponding polynomial equation becomes as Equation 10.7 (see case study in Chapter 10). Other examples of output as a function of inputs (Output $= f$ (Input) or $Y = f(x)$) are as follows:

In an *engineering* (R&D, manufacturing, and tool design) case the output and input variables are $Y_i =$ symptom, $x_i =$ cause
Y_i: Yield, waste, capacity, downtime, cycle time, etc.
x_i: Design (dimension, plastic types, material properties, etc.)
x_k: Material (chemical types, grade, amorphous, crystalline, shrinkage, melt index, etc.)

In *injection molding operation* the process input and output variables are as follows:
Y_i: Cycle time, yield, scrap, dimensions, quality, etc.
x_i: Process (pressure, temperature, speed, flow rate, etc.)
x_n: Mold (cooling, gate, runner, etc.)

In the same way, some of the other departments of an organization the variables affecting customer satisfaction and bottom line are cited as follows:

Finance
Y_1: Payroll
x_1: Manually checking on time cards regarding missing time or vacation (i.e., one did not punch in or state vacation)
Y_2: Time to distribute the finance package
x_2: Replacing monthly hard copies with electronic or digital, Internet system
Y_3: Accounts receivable (cash collection)
x_3: Reduction of cash collection time

Marketing
Y_1: Product (e.g., what is the need for marketing?)
x_1: Clear direction to engineering
Y_2: Placement (where it goes; where does it need to be placed?)
x_2: Specific target marketing
Y_3: Promotion (literature, training, how to use product)
x_3: Advertising, clear literature information

Sales
Y_1: Customer needs company assistance in learning how to use the product
Y_2: Train people (how to sell, how to troubleshoot, how to design for different applications)
x_1: Educate customers on the company profile
x_2: Thorough knowledge of product for training

Customer Service
Y_1: Accounts receivable aging (statement)
Y_2: Customers needs quicker response time after submitting their requests
x_1: Invoice lost in the mail
x_2: Incorrect pricing

Human Resources
Y_1: Turnover
x_1: Performance review timeliness, promotion, mentoring
Y_2: Benefits
x_2: Benchmark, overall surveys and suggestions

Product Distribution Center or Warehouse
Y_1: Inventory level (receipts and shipments)
x_1: Accuracy of SKUs and quantity
Y_2: Shipment method (weight of shipment, etc.)
x_2: Carrier handling, packing, stacking, and reduction transit time
Y_3: Delivery time
x_3: Choice of carrier, shipping on time, delivery appointment, and information services (or information technology)
Y_i: Downtime
x_i: Experience, troubleshooting techniques, and maintenance

Sigma

By definition *sigma* (σ), a Greek letter, is the statistical quality measurement of standard deviation from the mean. *Six Sigma* describes how a process performs quantitatively. In other words, it measures the variation of performance.

3.2 THE NORMAL DISTRIBUTION

Normal distributions are probability curves that have the same symmetric shapes. They are symmetric with data numbers more concentrated in the center than in the tails. The term *bell-shaped curve* is often used to describe normal distribution. The area under the curve is unity. The height of a normal distribution can be expressed mathematically in two parameters: of mean (μ) and the standard deviation (σ). The mean is a measure of center or location of average, and the standard deviation is a measure of spread. The mean can be any value from minus infinity to plus infinity (in between $\pm \infty$), and the standard deviation must be positive. Thus, the probability of $f(x)$ (see Equation 3.9) is equal to one. Suppose that x has a continuous distribution. Then for any given value of x, the function must meet the following criteria:

$f(x) \geq 0$ for any event x in the domain of f

Since the normal distribution curve (symmetric from the mean) meets the x-axis in the infinity (as shown in Figure 3.2), the area under the curve and above the x-axis is one. This can be calculated by integrating the probability density on a continuous interval from minus infinity to plus infinity (Equation 3.10).

$$\int_{-\infty}^{+\infty} f(x)dx = 1 = \text{Area under the normal distribution curve} \qquad (3.9)$$

Where the height of a normal curve (the normal density function) for random variable x is defined as

$$f(x) = \frac{1}{\sqrt{2\pi\sigma^2}} e^{\frac{-(x-\mu)^2}{2\sigma^2}} \quad \text{for x interval } (-\infty \leq x \leq +\infty) \qquad (3.10)$$

where $f(x)$ is the height of a normal distribution curve [$f(x) \geq 0$]; μ is the mean; π is the constant 3.14159; e is the base of natural logarithms, which is equal to

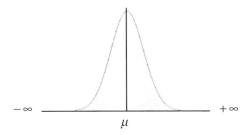

Figure 3.2   Normal distribution curve (symmetric) area equal unity.

2.718282; σ is the standard deviation of population; and $\dfrac{x - \mu}{\sigma} = z$ (this will be discussed in the z-distribution Section 3.3). The normal distribution curve for $n\sigma$ (n-sigma) is shown in Figure 3.3.

When n = 3, then for statistical quality control purposes, USL is equal to the mean (μ) plus three times the standard deviation ($\mu + 3\sigma$), and LSL is equal to the mean minus three times standard deviation ($\mu - 3\sigma$).

For any upper specification limit (USL) and lower specification limit (LSL) the probability density function (Equation 3.11) can be mathematically expressed as

$$A(x) = \int_{x=LSL}^{x=USL} \frac{1}{\sqrt{2\pi\sigma^2}} e^{\frac{-(x-\mu)^2}{2\sigma^2}} dx = \frac{1}{\sqrt{2\pi\sigma^2}} \int_{x=LSL}^{x=USL} e^{\frac{-(x-\mu)^2}{2\sigma^2}} dx$$

$$= \frac{1}{\sqrt{2\pi\sigma^2}} \int_{x=LSL}^{x=USL} e^{\frac{-\left(\frac{x-\mu}{\sigma}\right)^2}{2}} dx \qquad (3.11)$$

Thus, the probability (by definition probability is the area under the normal distribution curve A(x)) will be equal to 0.9973 when the process is centered on the target. This is the probability that 99.73% (Table 2.3) of data will fall within $\mu \pm 3\sigma$. This is also called 3σ capability, as shown in Equation 3.12.

$$A(x) = \int_{x=\mu-3\sigma}^{x=\mu+3\sigma} \frac{1}{\sqrt{2\pi\sigma^2}} e^{\frac{-(x-\mu)^2}{2\sigma^2}} dx = \frac{1}{\sqrt{2\pi\sigma^2}} \int_{x=\mu-3\sigma}^{x=\mu+3\sigma} e^{\frac{-(x-\mu)^2}{2\sigma^2}} dx = 0.9973 \qquad (3.12)$$

Now, by implementing the random variable z (standard z-transform or standard normal deviation), the new probability density function $f(z)$ is described as in the preceding section.

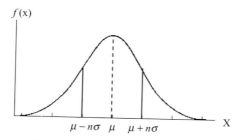

Figure 3.3 Normal distribution curve for $n\sigma$ (n-sigma).

3.3 THE STANDARD NORMAL DISTRIBUTION

The standard normal distribution (the z-distribution) is a normal distribution (see Equations 3.9–3.10) with a mean of zero ($\mu = 0$) and a standard deviation of one ($\sigma = 1$). Normal distributions can be transformed to standard normal distribution (z) by the expression

$$z = \frac{x - \mu}{\sigma} \qquad (3.13)$$

where x is a value from original normal distribution, μ is the mean of the original normal distribution, and σ is the standard deviation of the original normal distribution. If we replace x with USL in Equation 3.13, then the area under the normal curve beyond the USL will indicate the nonconformance probability above the specification limit. This is shown in Equation 3.14.

$$z = \frac{USL - \mu}{\sigma} \qquad (3.14)$$

As just mentioned, the standard normal distribution also is called the z-distribution. A z-value refers to the number of standard deviations (right or left) of the mean) from the mean for a particular score. For example, if a student scored an 85 (USL = 85) on the final test with a mean of 60 ($\mu = 60$) and a standard deviation of 10 ($\sigma = 10$), then using Equation 3.14 she scored 2.5 standard deviations above the mean. So a z-score of 2.5 means the original score was 2.5 standard deviations above the mean. The z-distribution will only be a normal distribution if the original distribution (x) is normal.

The Equation 3.13 will always yield a standard z-transform or standard normal deviation with a mean equal to zero ($\mu = 0$) and a standard deviation of one ($\sigma = 1$). Furthermore, the shape of the distribution will not be changed by the conversion. However, if x is not normal, then the standard z-transform will not be normal either.

Now, substituting the above information and Equation 3.13 in Equation 3.10 will result in Equation 3.15:

$$f(z) = \frac{1}{\sqrt{2\pi}} e^{\frac{-z^2}{2}}, \quad \text{where z is equal to } (-\infty < z < +\infty) \qquad (3.15)$$

The area within an interval $(-\infty, z)$ is given by Equation 3.16:

$$A(z) = \int_{-\infty}^{z} f(z) dz = \int_{-\infty}^{z} \frac{1}{\sqrt{2\pi}} e^{\frac{-z^2}{2}} dz = \frac{1}{\sqrt{2\pi}} \int_{-\infty}^{z} e^{\frac{-z^2}{2}} dz \qquad (3.16)$$

There is no need for integration of Equation 3.16. One can obtain the area of $A(z)$ for various z-numbers from the z-Table. (This table is available in the

Appendix.) The standard normal z-distribution curve for $z = \pm n$ is shown in Figure 3.4.

Example 3.1 shows the probability of defects for a given standard using standard normal distribution theory.

EXAMPLE 3.1.   What is the nonconformance probability when $z = 1.84$?

From z-Table (see Appendix) at 1.84 the area of standard normal distribution is equal to 0.4671 as shown in Figure 3.5. This represents the area between mean ($z = 0$) and $z = +1.84$ (area at $0 \leq z \leq 1.84$). One should keep in mind that the area in z-Table is designed for only one-half of the distribution curve.

Considering the two tails probability, the areas of the tails are

Area of right tail ≥ 1.84 equal to 0.0329
Similar the area of left tail ≤ -1.84 equal to 0.0329
The total area $= 2 \times 0.0329 = 0.0658$
Thus, the probability of defects $= 6.58\%$

The normal probability distribution (for sample numbers n $\geq$ 30) is the most common and popular in statistical analysis. However, there are other probability

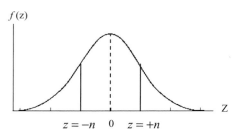

Figure 3.4   Standard normal z-distributions.

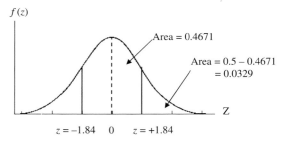

Figure 3.5   Area of distribution curve.

distributions (e.g., *t*-distribution), for sample numbers less than 30 where normal distribution will not give better results.

3.4 *t*-DISTRIBUTION

This is similar to a normal distribution (n ≥ 30) bell-shaped symmetrical curve, with the exception that it is used for small sample size (n < 30), thicker tails, and lesser height in the median frequency. Eventually, as sample size approaches 30 and higher, the distribution resembles a normal distribution. Other preceding properties differentiate *t*-distribution from normal distribution.

1. Applies to any sample size n < 30, the result is different for different sample sizes n.
2. Degrees of freedom are required.
3. The mean is zero ($\mu = 0$) the same as normal distribution.
4. Variance = $\lambda(\lambda - 2)^{-1}$, variance is larger than one, finally gets closer to one as sample size increases.
5. Range is $(-\infty, +\infty)$.
6. Distribution is symmetrical with respect to mean.
7. Population standard deviation is unknown.

Probability density function (pdf) for *t*-distribution is defined in Equation 3.17:

$$f(x) = G[0.5(\lambda + 1)]\left[(\lambda\pi)^{0.5} G(0.5\lambda)\right]^{-1} \left(1 + x^2\lambda^{-1}\right)^{-0.5(\lambda+1)} \qquad (3.17)$$

Where G is the Gamma function (the Gamma function extends the factorial function (n!)) and λ is the positive parameter called "degrees of freedom," as λ becomes larger, the *t*-distribution approaches normal distribution.

Since the *t*-distribution is normally used to establish a confidence interval and hypothesis tests, the preceding discussion concentrates on this concept for *t*-distribution testing.

3.4.1 CONFIDENCE INTERVAL FOR THE DIFFERENCE OF TWO MEANS

When testing large sample sizes (n ≥ 30) from two populations to compare the two means, the *z*-statistic in Equation 3.18 was used.

$$z = (\overline{x_1} - \overline{x_2})\left((s_1^2 n_1^{-1} + s_2^2 n_2^{-1})^{0.5}\right)^{-1} \qquad (3.18)$$

However, for small sample sizes (n_1 < 30 or n_2 < 30), Equation 3.18 is no longer valid to approximate standard normal. Thus, for small sample sizes, t-test is required for testing a confidence interval, as shown in Equation 3.19.

$$t = (\overline{x_1} - \overline{x_2})\left((s_1^2 n_1^{-1} + s_2^2 n_2^{-1})^{0.5}\right)^{-1} \tag{3.19}$$

A degree of freedom is needed for samples, assuming standard deviations are not equal as

$$df = \frac{(s_1^2 n_1^{-1} + s_2^2 n_2^{-1})^2}{\dfrac{(s_1^2 n_1^{-1})^2}{n_{11} - 1} + \dfrac{(s_2^2 n_2^{-1})^2}{n_2 - 1}} \tag{3.20}$$

in the case that standard deviations are equal $\sigma_1 = \sigma_2$, then Equation 3.20 becomes

$$df = n_1 + n_2 - 2$$

An approximate $(1 - \alpha) \times 100\%$ confidence interval for the difference of two mean $\mu_1 - \mu_2$ (in small sample) is

$$\overline{x_1} - \overline{x_2} - t_{0.5\alpha, df}\left(s_1^2 n_1^{-1} + s_2^2 n_2^{-1}\right)^{0.5} \text{ to } \overline{x_1} - \overline{x_2} + t_{0.5\alpha, df}\left(s_1^2 n_1^{-1} + s_2^2 n_2^{-1}\right)^{0.5} \tag{3.21}$$

EXAMPLE 3.2.　Independent random samples were chosen from two normal populations.

The following statistics were calculated.

Sample 1	Sample 2
$n_1 = 16$	$n_2 = 16$
$\overline{x}_1 = 3.50$	$\overline{x}_2 = 3.75$
$s_1 = 0.75$	$s_2 = 0.45$

A. Find a 90% confidence interval for $\mu_1 - \mu_2$, not assuming that population variances are equal $(\sigma_1 - \sigma_2)$.

$$df = \frac{\left((0.75)^2 (16)^{-1} + (0.45)^2 (16)^{-1}\right)^2}{\dfrac{\left((0.75)^2 (16)^{-1}\right)^2}{16 - 1} + \dfrac{\left((0.45)^2 (16)^{-1}\right)^2}{16 - 1}} = \frac{(0.0478)^2}{(8.240E - 5) + (1.068E - 5)} = 24.55$$

Rounding df to the nearest integer, $df = 25$. Using the t-Table in the Appendix, $t_{(0.10/2),25} = t_{0.05,25} = 1.708$.

The result of 90% confidence interval for $\mu_1 - \mu_2$ is

$$\overline{x}_1 - \overline{x}_2 - t_{0.05,25}\left(s_1^2 n_1^{-1} + s_2^2 n_2^{-1}\right)^{0.5} \text{ to } \overline{x}_1 - \overline{x}_2 + t_{0.05,25}\left(s_1^2 n_1^{-1} + s_2^2 n_2^{-1}\right)^{0.5}$$

$$= (3.50 - 3.75) - 1.708\left((0.75)^2 (16)^{-1} + (0.45)^2 (16)^{-1}\right)^{0.5} \text{ to}$$

$$(3.50 - 3.75) + 1.708\left((0.75)^2 (16)^{-1} + (0.45)^2 (16)^{-1}\right)^{0.5}$$

$$= -0.25 - 0.373 \text{ to } -0.25 + 0.373$$

$$= -0.623 \text{ to } +0.123$$

B. Find a 90% confidence interval for $\mu_1 - \mu_2$ using the estimate of variance for the populations.

Since sample variances are equal, say $\sigma_1^2 = \sigma_2^2 = \sigma^2$. Consequently, the combined variances give Equation 3.22:

$$s_c^2 = \left((n_1 - 1)s_1^2 + (n_2 - 1)s_2^2\right)(n_1 + n_2 - 2)^{-1}$$

$$s_c^2 = \left((16 - 1)(0.75)^2 + (16 - 1)(0.45)^2\right)(16 + 16 - 2)^{-1} \qquad (3.22)$$

$$s_c^2 = (8.4375 + 3.0375)(0.0333) = 0.3825$$

$$s_c = 0.62$$

If $s_c = 0.62$ is between $s_1 = 0.75$ and $s_2 = 0.45$, then, $s_c^2 = 0.62$ is an estimate of variance σ^2 of two (sample 1 and sample 2) populations.

To find a 90% confidence interval for $\mu_1 - \mu_2$, use Equation 3.23 or 3.24:

$$\overline{x}_1 - \overline{x}_2 - t_{0.5\alpha,df}\left(s_c^2 n_1^{-1} + s_c^2 n_2^{-1}\right)^{0.5} \text{ to}$$

$$\overline{x}_1 - \overline{x}_2 + t_{0.5\alpha,df}\left(s_c^2 n_1^{-1} + s_c^2 n_2^{-1}\right)^{0.5} \qquad (3.23)$$

or

$$\overline{x}_1 - \overline{x}_2 - t_{0.5\alpha,df}\left(s_c\right)\left(n_1^{-1} + n_2^{-1}\right)^{0.5} \text{ to}$$

$$\overline{x}_1 - \overline{x}_2 + t_{0.5\alpha,df}\left(s_c\right)\left(n_1^{-1} + n_2^{-1}\right)^{0.5} \qquad (3.24)$$

where $df = n_1 + n_2 - 2$ and $\alpha = 0.10$.

Using $df = n_1 + n_2 - 2 = 16 + 16 - 2 = 30$ and $\alpha = 0.10$ from the t-Table in the Appendix, $t_{(0.10/2),25} = t_{0.05,30} = 1.697$. Next the outcome of the confidence interval is

$$(3.50 - 3.75) - 1.697(0.62)\left(16^{-1} + 16^{-1}\right)^{0.5}$$

$$\text{to } (3.50 - 3.75) + 1.697(0.62)\left(16^{-1} + 16^{-1}\right)^{0.5}$$

or

$$(-0.25 - 0.372) \text{ to } (-0.25 + 0.371)$$
$$-0.622 \text{ to } 0.121$$

Comparing A and B results of the confidence interval, there is not much difference. However, more frequently these confidence intervals can have a larger difference between the two, depending on the sample sizes (n_1 and n_2) and variances (s_1^2 and s_2^2).

Hypothesis Testing for μ_1 and μ_2

For the two-tailed test: The goal is to test for H_0: $\mu_1 = \mu_2$ versus H_a: $\mu_1 \neq \mu_2$. Thus, reject H_0 if $|t| > t_{0.5\alpha, df}$
where from the t-Table, the value is $t_{0.5\alpha, df} = t_{0.05, 30} = 1.697$.
Now, assuming that $\sigma_1 = \sigma_2$, then t is equal to Equation 3.25:

$$t = \left(\overline{x_1} - \overline{x_2}\right)\left(s_c \left(n_1^{-1} + n_2^{-1}\right)^{0.5}\right)^{-1} \tag{3.25}$$

or

$$t = (3.50 - 3.75)(0.62(16^{-1} + 16^{-1})^{0.5})^{-1} = (-0.25)(0.22)^{-1} = -1.14$$

Because $|-1.14| = 1.14 < 1.697$, it therefore fails to reject H_0.

3.5 BINOMIAL DISTRIBUTION

It is a discrete distribution that has only (absolutely) two possible outcomes—for examples, good or bad, on or off, success or fail, and high or low. The probabilities are the same from experiment to experiment. Thus, it is used to achieve probability of x success (or defects) in m independent repetitions (or trials), with probability of success on a single experiment represented by or equal to p. Then, the given binomial probability mass function is defined in Equation 3.26:

$$P(x, p, m) = m![x!(m-x)!]^{-1} p^x (1-p)^{m-x} \tag{3.26}$$

where
$x = 0, 1, 2, \ldots, n - 2, n - 1, n$ (number of occurrences with success or fail)
m = number of repetitions
p = probability of success or fail
$!$ = called factorial (is the product of multiplying all the numbers beginning from 1 to a specified number) for instance,
$3! = (1)(2)(3) = 6$

$5! = (1)(2)(3)(4)(5) = 120$
$n! = 1 \times 2 \times 3 \times \ldots \times n$

Other properties are
Mean $= \mu = mp$
Variance $= \sigma^2 = mp(1 - p)$
Range $= 0$ to M
Standard Deviation $[mp(1 - p)]^{\frac{1}{2}}$

Additional properties are

1. For p = 0.5, the binomial distribution shape looks normal (bell-shaped).
2. For p < 0.5, the distribution curve is skewed right, and the skewness increases as p decreases.
3. For p > 0.5, the distribution curve is skewed left, and the skewness increases as p increases and nears 1.0.

However, as the sample number increases, the distribution approaches normal (symmetrical shape) no matter what the p value is.

EXAMPLE 3.3.   Suppose in a basketball game a player was fouled out during a 2-point basket. He is then asked to shoot a 2-point basket. What is the probability of the ball going through the basket?

The probability of the ball going through the basket is as follows:

Let's say basket = success, no basket = fail

Outcome	First Try	Second Try
1	Success	Success
2	Success	Fail
3	Fail	Success
4	Fail	Fail

Using Equation 3.26, the probabilities are

$$P(0, 0.5, 2) = 2![0!(2-0)!]^{-1}\, 0.5^0\, (1-0.5)^{2-0} = 0.25$$
$$P(1, 0.5, 2) = 2![1!(2-1)!]^{-1}\, 0.5^0\, (1-0.5)^{2-1} = 0.50$$
$$P(2, 0.5, 2) = 2![2!(2-2)!]^{-1}\, 0.5^2\, (1-0.5)^{2-2} = 0.25$$

The probabilities of the ball getting 0, 1, and 2 baskets are as follows:

Number of Baskets	Probability
0	0.25
1	0.50
2	0.25

3.6 POISSON DISTRIBUTION

Poisson distribution is a discrete distribution in terms of manufacturing environment and counts the number of times a product defect occurs over the specified volume, time, area, and length. Thus, for x to be a Poisson random variable, it has to be statistically independent in that the occurrence has to happen separately at different times. For example, an occurrence might happen between 1:00 AM and 4:00 AM and another between 4:00 PM and 6:00 PM.

1. The occurrence happens randomly in time or space, not in groups.
2. Normally it applies to processes that have a probability of defects less than or equal to 10%.
3. Mean = μ.
4. Variance = μ.

The Poisson probability of mass function, or the equation-calculating probability for Poisson distribution, is given in Equation 3.27:

$$p(x, \theta) = e^{-\theta}\theta^x (x!)^{-1} \text{ for } x = 1, 2, 3, \ldots \tag{3.27}$$

or

$$p(x) = e^{-\theta}\theta^x (x!)^{-1}$$

where

$p(x)$ = probability of x occurrences
θ = the expected number of event occurrences over the specified period of time or space interval, and is equal to mp
m = sample numbers
p = proportion defective (a constant, i.e., 5%)
$!$ = (called factorial) the product of multiplying all the numbers beginning from 1 to a specified number—for instance, $3! = (1)(2)(3) = 6$
e = 2.718281828

EXAMPLE 3.4. A quality assurance engineer was assigned to inspect a section (area) of a product. The engineer found that the average defect percentage of the product is about 3%, using sample sizes (*m*) of 150 products. Find the probability of discovering only *x* = 4 defectives in the sample products.

Using Equation 3.27:

$$p(x) = e^{-\theta}\theta^x (x!)^{-1}$$
$$= e^{-mp}(mp)^x (x!)^{-1}$$
$$= e^{-(150 \times 0.03)}(150 \times 0.03)^4 (4!)^{-1}$$
$$= (0.0111090)(17.086)$$
$$= 0.1898 \text{ or } 18.98\%$$

So there is an 18.98% chance of finding 4 defectives in the 150 sample parts.

3.7 EXPONENTIAL DISTRIBUTION

The exponential distribution is a continuous distribution that is extensively used in reliability engineering and estimation (Section 3.10) to describe the lifetime failure of a certain product or component of a machine (or system) in a specified period of time. If the random variable *t* represents time, α the mean time between failure, and $\beta = \alpha^{-1}$ the failure rate of the system, then the probability density function of the failure rate for exponential distribution can be written as $P_F(t) = 1 - e^{-\beta t}$

EXAMPLE 3.5.   Suppose a notebook computer battery has a usage life defined by exponential distribution with a mean of 25 hours. Find the probability that a battery will fail before its expected lifecycle of $\alpha = 25$ hours.

The battery lifecycle (*t*) has an exponential distribution with *t* = 25 and $\beta = \alpha^{-1} = (25)^{-1} = 0.04$.

$$P_F(t) = 1 - e^{-\beta t} = 1 - e^{-(25^{-1})(25)} = 0.632 = 63.20\%$$

Thus, there is 63.20% chance that the battery will fail.

3.8 HYPERGEOMETRIC DISTRIBUTION

The properties that apply to hypergeometric distribution and make it different than Poisson or binomial are as follows:

1. Discrete (discontinue with respect to time) processes
2. Small sample size or lots

3. Sampling with no replacement
4. Processes that number of defects are known.

Thus, the probability of mass function (PMF) for hypergeometric distribution for random variables is given in Equation 3.28:

$$p(x) = C_x^k C_{n-x}^{N-k} \left(C_n^N \right)^{-1} \tag{3.28}$$

where
$p(x)$ = probability of discovering x defects
n = sample numbers
N = population size
K = occurrence in the population
C_x^k = $k!(x!(k-x)!)^{-1}$ combination
C_{n-x}^{N-k} = $(N-n)!\{(n-x)![(N-k)-(n-x)]!\}^{-1}$
$(C_n^N)^{-1}$ = $[N!(n!(N-n))^{-1}]^{-1}$

EXAMPLE 3.6.   A young, growing company is making products in small lots. An inspector is assigned to do sampling from a particular manufacturing process. So the inspector takes a sample size of $n = 5$ from a lot size of $N = 18$ parts, where $k = 4$ occurrence in the population. Find the probability of only $x = 1$ defects in the sample.

Using Equation 3.28, the probability is

$p(x) = C_x^k C_{n-x}^{N-k} \left(C_x^N \right)^{-1}$
$n = 5$
$N = 18$
$K = 4$
$C_x^k = C_1^4 = 4!(1!(4-1)!)^{-1} = 24(6)^{-1} = 4$
$C_{n-x}^{N-k} = C_{5-1}^{18-4} = C_4^{14} = 14!(4!(14-4)!)^{-1} = (8.718E10)(1.10E-8) = 1001$
$\left(C_n^N \right)^{-1} = \left(C_5^{18} \right)^{-1} = \left[18!(5!(18-5)!)^{-1} \right]^{-1} = 0.000037973 = 3.7973E-5$
$p(1) = (4)(1001)(3.7973E-5) = 0.1520 \text{ (or } 15.20\%)$

3.9 NORMALITY TESTS

Kurtosis and Anderson Darling are two important normality tests.

3.9.1 KURTOSIS

Kurtosis is a factor that measures the peak or flat area of a bell-shaped probability distribution curve. Consider the two probability density functions (PDF) in Figure 3.6. It is difficult to estimate which distribution has a larger standard deviation. As a matter of fact, it is impossible. By a glance one might assume that the distribution on the right has a lower standard deviation due to the higher peak at the target. At the same time, one might think it has a higher standard deviation due to flatter tails.

The various shapes of the two distribution curves represent Kurtosis. Thus, Kurtosis is based on the size of the distribution's tails. In Figure 3.6 distribution on the right-hand side has larger Kurtosis than the one on the left. Sample Kurtosis can be calculated using Equation 3.29.

$$\text{Kurtosis} = \frac{\sum_{i=1}^{n}(x_i - \mu)^4}{n\sigma^4} - 3 \qquad (3.29)$$

where μ is mean and σ is the standard deviation of x_i.

Note that with the higher distribution, the area of the curve is still equal to 1.0.

3.9.2 ANDERSON DARLING

A normality test is a normal probability plot that is used to conduct a statistical test—that is, a hypothesis to find out if the distribution curve is normal or not. A best straight line (normal probability paper—log paper) represents a cumulative distribution for a sample/population from which data have been collected. After distribution data have been plotted for normality, one may also test for p-value. If p-value < 0.05, then it is not normal. However, if p-value ≥ 0.05, then

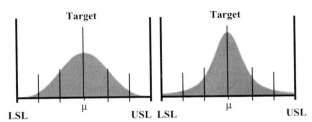

Figure 3.6   These curves represent the conception of Kurtosis. The PDF on the right has higher Kurtosis than the PDF on the left. It is more peaked at the center, and it has flatter tails.

the distribution curve is normal. Similarly, if p-value > 0.05, then the hypothesis test fails to reject.

3.10 RELIABILITY ENGINEERING AND ESTIMATION

Reliability (state of being dependable) engineering methodology is the application of a variety of techniques and reliability principles to the lifecycle of goods (or product lifetime) and equipment for testing and improving the product. Thus, reliability (also known as survival function) is defined as the probability of a defect-free product, operation, or service per a calculated period of time. In other words, successfully lasting life (i.e., functionality life) of a product is equal to or greater than a predetermined period of time. Reliability is quantified by the following three principles and techniques:

A. The **reliability function** $P_R(t)$ (probability of defect-free product or operation) is given in Equation 3.30.

$$\log_e P_R(t) = -\frac{t}{a} = -\beta t \tag{3.30}$$

or

$$P_R(t) = e^{-\frac{1}{a}} = e^{-\beta t}$$

where $\beta = \dfrac{1}{\alpha}$ and

$P_R(t)$ = The reliability function or probability of a defect-free operation for a time period t or beyond means no failure within the interval time—in other words no failure will occur until time t.

t = Specified period of defect-free operation.

α = Mean time between failures (MTBF), when the system is repairable.

β = Constant failure rate of the system (defect rate).

and further $\dfrac{1}{\beta}$ is called mean time to failure (MTTF) when the system is nonrepairable.

What Equation 3.18 refers to is that the lifecycle of any product under study is presumed to begin at time $t = 0$, and any problems occurring with respect to functionality are irrelevant before the time $t \le t_{specified}$ or time interval $(0, t)$. Since reliability is time-based quality, it is also a function of other factors that impact the time. In Equation 3.31

$$P_R(x) = f(x_1, x_2, x_3, \ldots, x_{n-2}, x_{n-1}, x_n) \tag{3.31}$$

where, $x_1, x_2, x_3, \ldots, x_{n-2}, x_{n-1}, x_n$ are the factors that impact "time-based" qualities of the system. To achieve product improvement objectives, reliability engineering has to be accurately defined, measured, analyzed, tested, verified, controlled, and maintained in the field or the system.

EXAMPLE 3.7. Suppose a product has shown 6,000 hours (250 days of production hours) of mean time between failures. Assuming a constant defect rate, what is the probability of a defect-free operation over a 24-hour interval (one day of operation)?

Assuming the reliability of an element (or product) exponentially dies (decays) through time t, as shown in Equation 3.30, the equation can be stated in the form of an exponential given in Equation 3.32.

$$P_R(t) = e^{-\frac{1}{a}} = e^{-\beta t} \tag{3.32}$$

$$\beta = \frac{1}{6,000} = 0.000167 = 1.67 \times 10^{-4} = 1.67E - 4$$

$$P_R(t) = e^{-\frac{24}{6,000}} = e^{-(24)(1.67\times10^{-4})} = 0.996008$$

Thus, the operation is 99.601% defect free over a one-day interval.

B. The **probability function** of failure time (or probability density function) is defined as the failure time distribution function $P_F(t)$, which is a supplement of the reliability function $P_R(t)$. In other words, the sum of both probabilities is equal to 1.0 (100%). That is,

$$P_R(t) + P_F(t) = 1.0 \tag{3.33}$$

Equation 3.33 can be expressed as

$$P_R(t) = 1 - P_F(t) \tag{3.34}$$

or

$$P_F(t) = 1 - P_R(t) = 1 - e^{-\beta t}$$

By taking the derivative of Equation 3.33, one can obtain the probability of the density function of failure, as shown in Equation 3.35.

$$P_f(t) = \beta e^{-\beta t} \tag{3.35}$$

Where $P_f(t)$ is the failure function, β is the failure rate, and t is the duration of a product lifetime, some values of $e^{-\eta}$ are given in Table 3.1.

C. The **hazard** (or failure) **rate function** $P_H(t)$ is defined as the limit of "failure rate $(P_H(t))$" as the time interval $(\Delta t \rightarrow 0)$ approaches zero.

$$P_H(t) = \frac{P_f(t)}{P_R(t)} = \frac{P_f(t)}{1 - P_F(t)} = \frac{\beta e^{-\beta t}}{1 - (1 - e^{-\beta t})} = \frac{\beta e^{-\beta t}}{e^{-\beta t}} = \beta \tag{3.36}$$

Table 3.1

Values of $e^{-\eta}$

η	$e^{-\eta}$	η	$e^{-\eta}$	η	$e^{-\eta}$	η	$e^{-\eta}$	η	$e^{-\eta}$
0.0000	1.0000	0.01	0.9900	1.10	0.3329	3.10	0.0450	5.10	0.0061
0.0001	0.9999	0.02	0.9802	1.20	0.3012	3.20	0.0408	5.20	0.0055
0.0002	0.9998	0.03	0.9704	1.30	0.2725	3.30	0.0369	5.30	0.0050
0.0003	0.9997	0.04	0.9608	1.40	0.2466	3.40	0.0333	5.40	0.0045
0.0004	0.9996	0.05	0.9512	1.50	0.2231	3.50	0.0302	5.50	0.0041
0.0005	0.9995	0.06	0.9418	1.60	0.2019	3.60	0.0273	5.60	0.0037
0.0006	0.9994	0.07	0.9323	1.70	0.1826	3.70	0.0247	5.70	0.0033
0.0007	0.9993	0.08	0.9231	1.80	0.1653	3.80	0.0224	5.80	0.0030
0.0008	0.9992	0.09	0.9139	1.90	0.1496	3.90	0.0202	5.90	0.0027
0.0009	0.9991	0.10	0.9048	2.00	0.1353	4.00	0.0183	6.00	0.0028
0.001	0.9990	0.20	0.8187	2.10	0.1225	4.10	0.0166	6.10	0.0022
0.002	0.9980	0.30	0.7408	2.20	0.1108	4.20	0.0150	6.20	0.0020
0.003	0.9970	0.40	0.6703	2.30	0.1002	4.30	0.0136	6.30	0.0018
0.004	0.9960	0.50	0.6065	2.40	0.0907	4.40	0.0123	6.40	0.0017
0.005	0.9950	0.60	0.5488	2.50	0.0821	4.50	0.0111	6.50	0.0015
0.006	0.9940	0.70	0.4966	2.60	0.0724	4.60	0.0101	6.60	0.0014
0.007	0.9930	0.80	0.4493	2.70	0.0672	4.70	0.0091	6.70	0.0012
0.008	0.9920	0.90	0.4066	2.80	0.0608	4.80	0.0082	6.80	0.0011
0.009	0.9910	1.00	0.3679	2.90	0.0550	4.90	0.0074	6.90	0.0010
				3.00	0.0498	5.00	0.0067	7.00	0.0009

where, $P_H(t)$ is the hazard rate function, and β is the positive constant failure. Equations 3.35 and 3.36 can also be written as in Equation 3.37.

$$P_f(t) = P_R(t) \times P_H(t) = \beta e^{-\beta t} \qquad (3.37)$$

EXAMPLE 3.8.   What is the reliability of a lightbulb at 2,500 hours, if the target service of the bulb was 1,600 hours for 4,100 bulbs and the total number of failures was 250?

Using an exponential distribution Equation 3.20,

$$\beta = \frac{1}{n\alpha} = \frac{1}{(1,600)(4,100)} = 0.0000001524 = 1.524E - 7$$

$$P_R(t) = e^{-\frac{1}{\alpha}} = e^{-\beta t}$$

$$P_R(2,500) = e^{-(1.524E-7)(2500)} \cong 0.99962 = 99.96\%$$

The following are other factors of time-based qualities:

- Maintainability
- Serviceability issues—that is, product-related equipment
- Repairability problems
- Availability
- Durability
- Safety and health-related issues

3.11 QUALITY COST

In general, the cost of quality can be estimated using Equation 3.38 as follows:

$$Y_{qc} = \frac{(X_{md} + X_{ld} + X_{rc}) + (X_{dpc} + X_{ac})}{X_{cmc}} \tag{3.38}$$

where
Y_{qc} = Cost of quality
X_{md} = Material defect
X_{ld} = Labor defect
X_{rc} = Rework cost (cost of correcting defects and conforming them to customer specification limits)
X_{dpc} = Defect prevention cost
X_{ac} = Appraisal cost
X_{cmc} = Complete manufacturing cost

The summary of quality cost categories and subcategories is as follows:

The Internal Defects

- Rework, cost of correcting defects and conform them to the customer specification limits
- Material, labor, and overhead defects on unfixable components
- Failure analysis

The External Defects

- Warranty returns
- Customer complaint correction
- Material returns (receipt and replacement)
- Cost to customer

Appraisal Cost

- Testing instruments (capital and maintaining)
- Inspection
- Product quality audit

Defect Prevention Cost

- Process quality control
- Planning and training

Chapter 4

Six Sigma Continuous Improvement

"The significant problems we face cannot be solved at the same level of thinking we were at when we created them."

—Albert Einstein

4.1 *SIX SIGMA* CONTINUOUS IMPROVEMENT PRINCIPLES

Advancement to near perfection (maximum profitability) is virtually impossible without integration of proper engineering design, material, process, and control strategies—in other words, achieving higher standard deviations in existing and future production processes. *Six Sigma* tools uncover the unseen root causes of potential problems and attack them to eliminate defect opportunities. This means that *Six Sigma* takes the necessary measurements at the early stages of product or process development before the problem occurs. Alternatively, it will focus on the processes indicating that their sigma level is either too low and cannot improve or a high sigma level (5 or more), which is too challenging to improve. One should keep in mind that the cost of redesign or correcting aftermath problems is extremely high and costly. (See Chapter 8 for more details.)

Six Sigma can be achieved only with cross-functional groups or joint efforts throughout the organization with great intensity—that is, if it is required by corporate culture and deployed firmly. Indeed, *Six Sigma* has to be applied from raw material all the way to the finished product and the shipping department.

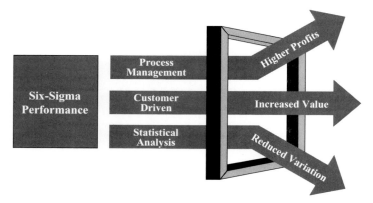

Figure 4.1   *Six Sigma* breakthrough performance model.

The performance of this is shown in Figure 4.1, the *Six Sigma* breakthrough model, with three main objectives: higher profit, increased value, and reduced variation. *Six Sigma* science of continuous improvement concentrates on two processes that each include three steps:

Process Characterization

1. Define project and process measurement (diagnosis)
2. Evaluate existing sigma (capability study)
3. Analyze process data

Process Optimization/Simulation

4. Improve and optimize process
5. Evaluate new sigma (capability study)
6. Control and maintain the process

Or the above steps can be summarized in DMAIC (define/measure [diagnosis], analyze, improve [optimize], control).

4.2 *SIX SIGMA* SYSTEMS

Six Sigma systems include the following concepts:

1. The fastest rate of improvement in customer satisfaction, cost, quality, process speed, and invested capital.
2. A business improvement and growth system, which leads to a new level of performance.

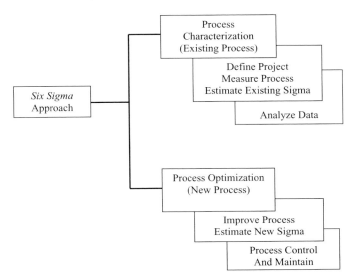

Figure 4.2   Representation of *Six Sigma* define and measure, analyze, improve or optimize, and control (DMAIC) levels.

3. A systematic data-driven approach to analyzing the root cause of manufacturing, as well as business problems/processes and dramatically improving them.
4. Improved decisions based on knowledge and data.
5. A financial results-driven system ($ performance paramount).
6. A project-driven system based on customer needs.
7. A system applicable to all parts of a business.
8. Aimed at the problem where the solution is not known.
9. A system with proven performance.
10. Improve customer satisfaction and supplier relationships.
11. Expand knowledge of product and processes.
12. Develop a common set of tools and improvement techniques.

Figure 4.2 illustrates the graphical representation of *Six Sigma* improvement steps.

4.3 *SIX SIGMA* IMPROVEMENT AND TRAINING MODELS

One may obtain *Six Sigma* training certification by completing the improvement models for Green Belt and Black Belt. These trainings are available through academic institutions, as well as quality societies or other certified organizations.

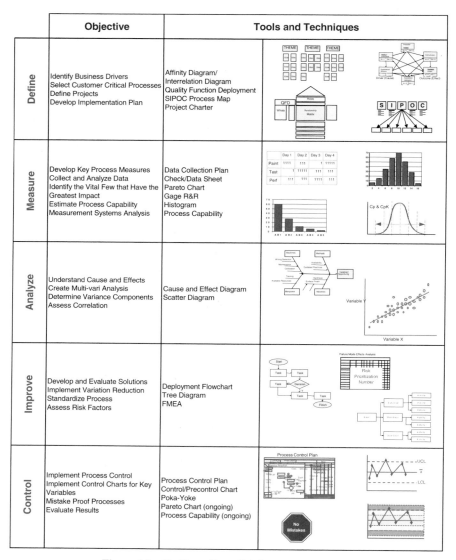

	Objective	Tools and Techniques
Define	Identify Business Drivers Select Customer Critical Processes Define Projects Develop Implementation Plan	Affinity Diagram/ Interrelation Diagram Quality Function Deployment SIPOC Process Map Project Charter	
Measure	Develop Key Process Measures Collect and Analyze Data Identify the Vital Few that Have the Greatest Impact Estimate Process Capability Measurement Systems Analysis	Data Collection Plan Check/Data Sheet Pareto Chart Gage R&R Histogram Process Capability	
Analyze	Understand Cause and Effects Create Multi-vari Analysis Determine Variance Components Assess Correlation	Cause and Effect Diagram Scatter Diagram	
Improve	Develop and Evaluate Solutions Implement Variation Reduction Standardize Process Assess Risk Factors	Deployment Flowchart Tree Diagram FMEA	
Control	Implement Process Control Implement Control Charts for Key Variables Mistake Proof Processes Evaluate Results	Process Control Plan Control/Precontrol Chart Poka-Yoke Pareto Chart (ongoing) Process Capability (ongoing)	

Figure 4.3   *Six Sigma* Green Belt improvement model.

	Objective		Tools and Techniques
Define	Identify Business Drivers Select Customer Critical Processes Define Projects Develop Implementation Plan	Affinity Diagram/ Interrelation Diagram Quality Function Deployment SIPOC Process Map Project Charter	
Measure	Develop Key Process Measures Collect and Analyze Data Identify the Vital Few that Have the Greatest Impact Estimate Process Capability Measurement Systems Analysis	Data Collection Plan Check/Data Sheet Pareto Chart Gage R&R Histogram Process Capability	
Analyze	Understand Cause and Effects Create Multi-vari Analysis Determine Variance Components Assess Correlation	Cause and Effect Diagram Multi-vari Charts Scatter Diagram	
Improve	Develop and Evaluate Solutions Implement Variation Reduction Standardize Process Assess Risk Factors	Design of Experiments Deployment Flowchart Tree Diagram FMEA	
Control	Implement Process Control Implement Control Charts for Key Variables Mistake Proof Processes Evaluate Results	Process Control Plan Control/Precontrol Chart Poka-Yoke Pareto Chart (ongoing) Process Capability (ongoing)	

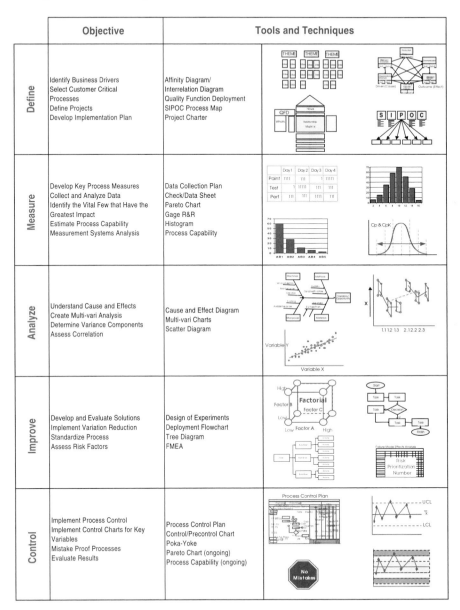

Figure 4.4 *Six Sigma* Black Belt improvement model.

On the other hand, you can apply all the procedures and guidelines of this book to any desired projects. Once you have knowledge and experience, you can participate in certified examinations through a quality organization. The training models for the Green Belt and Black Belt are cited in Figures 4.3 and 4.4. The details of these models are discussed in Chapter 8.

Chapter 5

Design for *Six Sigma*: Roadmap for Successful Corporate Goals

"Higher profit earnings by preventing problems rather than fixing them."

—Philip B. Crosby

5.1 DESIGN FOR *SIX SIGMA* (DFSS) PRINCIPLES

As mentioned in previous chapters, most of the quality problems (70 to 80%) are design related. K. Hockman's *Six Sigma Forum* (2001); N. Suh's *The Principles of Design* (1990); and R. Paul's *PDMA Handbook of New Product Development* (1996) all support that design issues should be solved at early stages of design rather than later. Problem solving (or process improvement) at the downstream end is more costly and time consuming than at the source.

Since *Six Sigma* is a business strategy, executive leadership is the formula for the success of *Six Sigma* projects. Basically, the achievement of significant results is based on the commitment and leadership of executive management. Figure 5.1 represents a companywide success model. The model illustrates that the probability of a company's success increases by selecting the right company's assets to lead the projects. Furthermore, the success of design for *Six Sigma* is dependent on the following criteria:

Customer Focus

1. Customer requirements: technical specification and measurements such as upper and lower specification limits (USL/LSL)

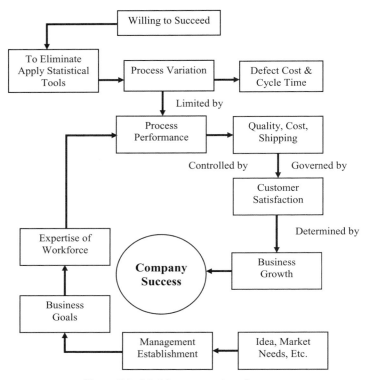

Figure 5.1   Model representation of success.

2. Voice of customer (VOC): what customers need and want
3. Critical to quality (CTQ): critical requirements of customers
4. *Six Sigma* design: designing and developing products and processes based on customer specification limits
5. Integrating design for *Six Sigma* (DFSS) concepts with the customer's design

Bottom-Line Focus
• As opposed to other programs that concentrate on quality, *Six Sigma* focuses on financial impact.

Top Management Commitment
• Management engagement and full support in the project

Today's Top Business Challenges
• To bring products to market faster than the competition
• Reduce company's cost
• Deliver high-quality products

Six Sigma **Staff Commitment**
All of the company's *Six Sigma* staff (e.g., leaders and management) should learn and have a commitment to (1) *Six Sigma* methods and tools, (2) completion of projects, (3) training and promoting the *Six Sigma* program, and, finally, (4) they must be positive thinkers with a "can-do attitude."

Infrastructure Implementation
The priorities are to build infrastructure, provide training, coordinate initiatives, track progress, and identify best practices in the market (benchmarking).

Engineering Robust Design
Experimental design is an efficient technique of experimentation that identifies key process input variables. It also identifies the optimum settings that affect the process mean and output variations with a minimum of testing, such as the following:
- Experimental design
- Taguchi's robust design
- Central composite design
- Response surface methodology
- Design validation
- Tolerance design
- Analysis of variance (ANOVA)

Pilot Experimentation
- Create a prototype based on the robust engineering design

Figure 5.2 illustrates the design for *Six Sigma* process and tools that is required to achieve DFSS goals.

5.2 DESIGN FOR *SIX SIGMA* STEPS

As shown in Figure 5.3, these are the DFSS steps:

- **Define** the project goals and customer (internal and external) requirements.
- **Measure** and determine customer needs and specifications (LSL and USL); benchmark competitors and similar industries.
- **Analyze** the process or product options to meet customer needs.
- **Design** (detailed) the process or product to meet customer needs.
- **Verify** the design performance and ability to meet customer needs.

In general, DFSS is a function of measurement and design variables (independent factors).

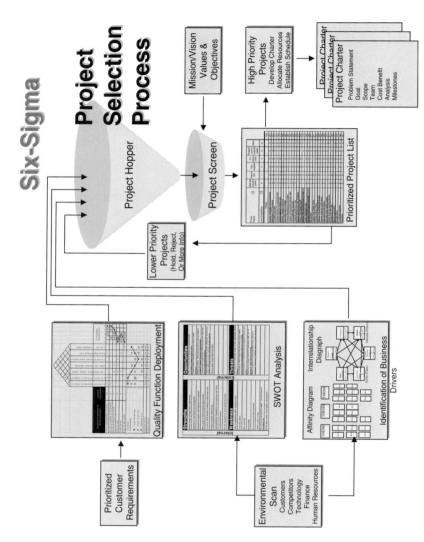

Figure 5.2 Project selection process.

52

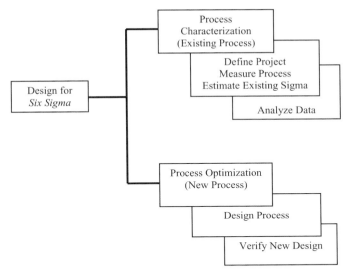

Figure 5.3 Representation of *Six Sigma* design DMADV (define/measure [diagnosis], analyze, design, verify) levels.

$$\text{DFSS} = f \text{ (measure variables, design variables)}$$

The following DFSS steps should be considered for data analysis:

1. Identify all the measurements in the form of mathematical numbers.
   In any process all the activities should be measured in numbers.
2. Specify the most dominant variables in the design.
   In any production processes there are variables that have a greater effect on process performance than others.
3. Determine variables that contribute to the sigma mean.
4. Identify the source of variation/root causes of defects.

When designing plastic molded parts, consideration should be given to aesthetics, functional process capabilities, and manufacturing constraints. All disciplines from product design to production phase should understand their roles in bringing the product to completion. There is an interrelationship between design, materials, and processes. Consequently, any decisions made by individuals and groups will profoundly effect functionality, process efficiency, tooling, assembly maintenance, engineering, and the total lifecycle of the product. Some of the factors that need to be considered are (1) design tolerance, (2) property consistency of material, (3) tooling capability, and (4) process capability.

5.3 *SIX SIGMA* ERGONOMICS

Ergonomics is an engineering applied science or sciences of designing a process/product that is based on how humans think, see, and behave. It is a manufacturing process or product that suits the needs of humans without any error, stress, or discomfort and creates a highly positive impact when used. Hence, application of ergonomics to a complex system converts it into a much simpler one without creating any problems. This results in a worry-free process or product. It is also called human factors engineering. Ergonomics engineers usually concentrate on process/product modeling and design, workforce performance and ability, work environmental space, type of machines, computational equipment, and so forth.

Ergonomics has its greatest effect when it is implemented from the very starting point of a new project. It includes two basic limitations of human performance: physical and perceptual/intuitive. A process or product designer should study the population—that is, in a manufacturing environment of the employee's size, gender, and age, along with his or her motions, strengths, and so forth. The collected information should help the designer to have a clear picture and ensure that the final process or product will physically match the employee's or consumer's needs. Perceptual behavior of human beings is dependent on gaining knowledge through the senses. On the other hand, intuitive behavior is based on perception, reasoning, and logic.

Following the product/process design layout, the ergonomic role should be implemented. Likewise, after the design phase and prototype (or preliminary) process has been established, evaluation procedures need to take place. Numerous research and surveys illustrate those companies that have adopted ergonomics concepts in their programs and have improved cost savings and profits through the quality of their products.

These are some of the benefits and highlights of product development in the process of design for *Six Sigma*:

1. Increase in productivity
2. Reduction in overall costs
3. Reduction in product development time and development costs
4. Reduction in defects and variability in the product development process (PDP)
5. Increase in profit over the product lifecycle
6. Increase in customer satisfaction

Moreover, DFSS will bring products to the market faster, take less time to achieve volume and quick profit, and reduce the time in delivering high-quality products. The "Factor-of-Ten Rules" in DFSS are the cost of fixing a defect based on the stage where it was found, as follows:

1. Cost of a problem found in the concept phase (very early stage) is equal to $10.
2. Likewise at feasibility phase = $100, and the others are design = $1,000; production = $10,000; and product in the market = $100,000.

5.4 TOOLS AND TECHNIQUES

The first few steps of tools and techniques used in DFSS are very similar to *Six Sigma* continuous improvement: define, measure, and analyze. The last two steps change to design and verify—basically, implementation of DMADV tools versus DMAIC *Six Sigma* roadmaps. The highlights are evaluation of new-product development process actions and product development performance measures.

Product development process actions are as follows:

1. Concept development phase
   * Technology concepts
     * Innovation, design, invention
   * Competitive capability benchmarking
   * Quality function deployment
   * Documentation
2. Design development phase
   * Functional analysis
   * Flow diagram
   * Mathematical modeling of capability, such as $y = f(x)$
   * Failure mode effects analysis (FMEA)
   * Tree analysis
3. Optimization development phase
   * Failure mode effects and analysis (FMEA)
   * Fault tree analysis (FTA)
   * Critical-to-function (CTF) specification
4. Verify capability and functionality phase
   * Tolerance design analysis
   * Capability assessment (C_p, C_{pk}, P_p, P_{pk})
5. Feasibility
6. Development
7. Preproduction
8. Production
9. Product in the market

And the benefits are as follows:

1. Increase in productivity
2. Reduction in costs

3. Reduction in product development time
4. Increase in product quality
5. Increase in profit over the product lifecycle
6. Increase in customer satisfaction

Value stream mapping and management, is a method of visually mapping a product's production path from "door-to-door." It also consists of all the functions and actions of the existing process required to bring a product through the main flows essential to the final product. (For more details see Chapter 6.)

The quality function deployment (QFD) is a relationship matrix of *how* the company measures and *what* are the customer's requirements. The failure mode effects analysis (FMEA) is a product risk assessment that analytically approaches the prevention of defects by prioritizing potential problems and their resolution. Additional tools are: design for concept, vision, and process layout; synchronizing product designs with process capabilities; and designing a project economically.

Poka-Yoke (a Japanese term for mistake proofing) is defined as a process of analysis and implementation to build quality into an assembly or manufacturing process with simple low-cost devices and methods such as:

• Identify problems
• Prioritize problems
• Find the root cause
• Create solutions
• Measure the results

Six Sigma is required to be designed for reliability, availability, and maintainability (see Section 3.5 for more details). It also should be designed for service operations, manufacture, and assembly. It will incorporate the appropriate design metrics and performance measurements, for example; making a product mistake proof results in zero defects. This is defined as a process of analysis and implementation to build quality into an assembly or manufacturing process with simple low-cost devices and methods. The output will illustrate a process design that is robust to mistakes.

A phased toll gate approach involves the following steps:

1. Get the team members involved.
2. Make sure that opportunities are clearly communicated.
3. Clearly define roles and responsibilities of team members.
4. Establish a project review with team members.
5. Review the project with the team as well as the sponsor and have the sponsor sign-off upon approval.

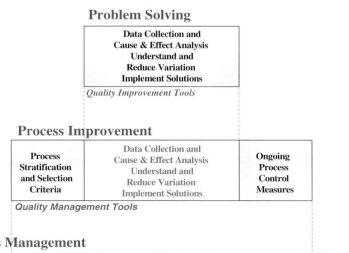

5.5 PROCESS MANAGEMENT

Process management is a combination of (1) problem-solving techniques, which use quality improvement tools to resolve issues; (2) process improvement that uses quality management tools to improve processes; and (3) process management that uses advanced quality planning tools to manage the process, as shown following.

Ultimately the process owner is responsible for the management and success of the project. Nevertheless, managing the process/project is one of the most difficult and challenging assignments in any organization. A good project manager will motivate and influence team members by making them feel equally important in the project and by ignoring the members' ranks, as well as educating team members about the importance of the project. Effective project managers have strong tasks and goals in order to keep the project on schedule and be able to meet the deadline, especially to motivate and obtain results from other team members on the projects.

Different Types of Processes

Identity

This is a process that defines the company for itself, its customers, and its
investors. It differentiates a firm from its competitors and is located at the
heart of the firm's success.

Priority

A priority process is one that determines the firm's effectiveness. It strongly
influences how well identity processes are carried out and how a firm
stands relative to its competition.

Background

Processes that are necessary to support daily operations. Many
administrative and overhead functions are background processes.

Mandated

These are processes that a company carries out only because it is legally
required to do so.

Chapter 6

Design for Lean/Kaizen
Six Sigma

Lean means speed and quick action (reducing unneeded wait time).
Six Sigma means identifying defects and eliminating them.
Lean *Six Sigma* Engineering means best-in-class. It creates value in the
  organization to benefit its customers.

6.1 LEAN *SIX SIGMA* AND PRINCIPLES

As we all know, time is money (it is also an investment), and money is the
bottom line for any business organization to grow and succeed. We also know
that defects are nothing but loss. So Lean *Six Sigma* is the solution that enables
companies to move ahead of the competition. With that in mind, Lean *Six Sigma*
combines the two most important improvement techniques in today's business
and industry. One is the epitome of quality, which uses *Six Sigma* strategies, and
the other is Lean principles in achieving world class performance (WCP). Lean
applies to all processes in all industries and organizations. Mathematically the
WCP is the output response defined as a function of Lean and *Six Sigma*. In other
words, WCP is dependent on Lean (y_L) and *Six Sigma* (y_{SS}). Thus, as a mathematical function:

World class performance =
f (Lean causes of variation, *Six Sigma* causes of variation)

The first round of measurements, according to Equation 3.3, is shown in Equation
6.1:

$$\text{Output} = f(\text{Input})$$

$$Y_{WCP}(or\ Y_{LSS}) = f(y_L, y_{SS}) = f[f_L(x), f_{SS}(x)]$$

$$y_L = f(y_1, y_2, y_3, \ldots, y_{n-2}, y_{n-1}, y_n) \tag{6.1}$$

$$y_{SS} = f(y_1, y_2, y_3, \ldots, y_{n-2}, y_{n-1}, y_n)$$

where Y_{WCP} is world class performance response and Y_{LSS} is Lean *Six Sigma* output. Y_{LSS} is a function of the second round of measurements y_L (Lean output) and y_{SS} (*Six Sigma* output). Notations $(y_1, y_2, y_3, \ldots, y_{n-2}, y_{n-1}, y_n)$ represent the results of design, process, wait time, and so on. The second round of measurements arises from causes of Lean *Six Sigma*—that is, overproduction, excess inventory, design issues, dimensional variation, process condition, and machine downtime, as indicated in Equation 6.2:

$$\begin{aligned}
y_{\text{overproduction}} &= f(x_1, x_2, x_3, \ldots, x_{n-2}, x_{n-1}, x_n) \\
y_{\text{wait-time}} &= f(x_1, x_2, x_3, \ldots, x_{n-2}, x_{n-1}, x_n) \\
y_{\text{design}} &= f(x_1, x_2, x_3, \ldots, x_{n-2}, x_{n-1}, x_n) \\
y_{\text{process}} &= f(x_1, x_2, x_3, \ldots, x_{n-2}, x_{n-1}, x_n)
\end{aligned} \tag{6.2}$$

where y is response or measured values—that is, symptoms of performance, production, manufacturing, and product functionality. f is a response function of y_L (Lean) and y_{SS} (*Six Sigma*) performance. $x_1, x_2, x_3, \ldots, x_{n-2}, x_{n-1}, x_n$ are causes of independent variables—that is, variations in wait time factors, design factors, process, and so on.

The process of the Lean *Six Sigma* model is shown in Figure 6.1. Details of the above concept are described in Section 6.5, including mathematical modeling of Lean *Six Sigma* and design of experiment (DOE). Throughout this chapter, Lean and *Six Sigma* measuring factors and principles will be cited in the form of functions, which will be discussed later in the chapter.

The following rules and principles are required when designing and processing for a Lean *Six Sigma* project to meet the world class performance criteria:

1. Apply *Six Sigma* concepts to define phase, measure, analyze, develop/improve, and verify/control (DMAIC). Likewise, *Six Sigma* is a function of *Six Sigma* process measurement and improvement variables:

   Six Sigma = f(measure variables, improve variables)

2. Concentrate on quality and cost.
3. Focus on the company's bottom line.
4. Confirm on-time delivery, which is the customer's priority.
5. Improve productivity.
6. Emphasize safety and environment.
7. Focus on the voice of customer.

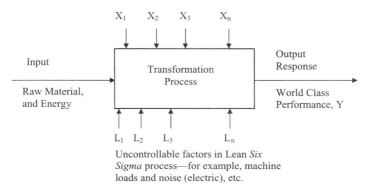

Figure 6.1   General model of Lean *Six Sigma* process.

- Match customer's specification limits.
- Identify customer's critical-to-quality (CTQ) issues and try to meet their requirements.
- Focus on elements of Lean manufacturing: faster, better, and cheaper.
- Meet customer satisfaction with speed and quality.
8. Focus on reduction of delay time—for example, 80% of the delay in any process is caused by 20% of the activities (workstations/time trap).
9. Focus on reduction of *wait* time—for example, 95% of lead time (the difference between the time an order is placed and when it is delivered by the supplier) in most processes is *wait* time. This exceeds by far the 5% of the time that is needed to get the job done, therefore slow processes are expensive processes, as shown below:

Lead time = Time an order is received − Time an order is placed
= (processing, shipping, and handling) time

10. Improve the slow processes that are expensive and costly.
11. Apply 5S improvement (Kaizen). (See Section 6.4 for details.)
12. Just-in-time (JIT)/Kanban. (See Section 6.4 for details.)
13. Total production maintenance (TPM).
    - Prolong productivity performance (PPP). (See Section 6.4 for details.)

6.1.1 ELEMENTS OF LEAN MANUFACTURING/PRODUCTION

The elements are as follows:

Reduction of Process Lead Times
Faster turnaround in any project is an important factor of lead time. It will also improve cycle time (Takt time—see Section 6.2.1), smaller lot sizes, and flexibility.

Improvement of Process and Product Quality to Zero Defects
This will result in *better quality*. It will also improve customer perception and satisfaction.

Minimization of Cost
This change will result in a *cheaper and more competitive price*. Thus, time is the single most important factor of competitiveness.

Inventory Reduction
Inventory is another important factor of Lean manufacturing—for instance, raw materials, work-in-process (WIP), and finished goods.

The preceding statements support the idea of time as the single best indicator of competitiveness. Additional principles of Lean *Six Sigma* and its benefits are listed in Table 6.1.

Other benefits of Lean manufacturing in both improvement and reduction include the following:

1. Improvement opportunities in the manufacturing system, such as product quality, productivity, customer service, capacity, inventory, WIP transportation in the system, flexibility, and process standardization (implementing the best process/operation techniques in completing project activities or tasks).
2. Reduction opportunities in the system of manufacturing consist of inventory, lot sizes, unit costs, design time, floor space, energy usage, and lead times (reduction of lead time and lead time variation to more consistent).

Table 6.1

Principles of Lean *Six Sigma*

Business process cycle time	50 to 90% reduction
Manufacturing cycle time	40 to 95% reduction
Inventory	40 to 80% reduction
Manufacturing floor space/office area	30 to 60% reduction
Productivity	25 to 60% improvement
New product development lead time	30 to 50% improvement
Manufacturing/operating costs	15 to 25% reduction
Cost of poor quality	30 to 50% reduction
Service costs	30 to 60% reduction
Service delivery time	30 to 50% improvement
Capacity	15 to 20% expansion

6.1.2 WASTE TYPES IN LEAN MANUFACTURING

Lean philosophy also recognizes the seven most important types of wastes. Moreover, manufacturing waste is also a function of the same variables:

Manufacturing waste = f (overproduction, wait time, transportation, processing, inventory, excess motions, and defects)

Overproduction

Parts are being produced without any new order or demand from the customer. Excess products may be sold with reduced prices at the end of the industry fiscal year to match the budget or lower the inventory for the new year's production. Here are some brainstorming questions related to overproductions:

- Why are you producing more than is requested by the customer?
- What storage problems and costs does it cause?
- Are you producing simply because you can—that is, because you have extra time and resources?
- How does this affect the line downstream?

Thus, overproduction waste is a function of

Overproduction = f (lower pricing, interest payment on loans, energy, extra manpower, and more)

Delay and Wait Time

Some common wait time is caused by processing delays, machine or system downtime, response time, or signature required for approval wait time. One may consider the question: How much time value could be added if wait time was transformed into beneficial or work time? The answer is: A huge amount! Likewise, here are some wait time value-added questions:

- Why is the delay happening?
- Are you waiting for materials, the next machine to be ready, or extra help to complete the job?
- What needs to change to make the flow smooth and even?

So, wait time depends on

Wait time = f (machine downtime, response time, signature approval, etc.)

Transportation

Transportation is defined as delivering to and from outside the factory warehouse facility. The transportation of finished goods normally is generated by poor plant process or unnecessary plant process layouts. Here are some questions relating to transportation:

- Is all the current transport of materials really necessary?
- How far have materials or parts traveled from the previous process?
- How far to the next process?
- Can better process layout and/or storage solutions reduce your transport time and distance?

Thus, transportation is a function of

$$\text{Transportation} = f(\text{plant process layout, travel distance, etc.})$$

Processing and Complexity

Storing work-in-process (WIP) products in further locations adds unneeded processing steps to complete the project and more. Here are some brainstorming questions:

- Do you lose efficiency due to poor functioning tools and equipment?
- Do you need cleaning and repair, or does process require redesign?
- How much processing is done that is overkill by, for example, overgluing, polishing parts that are never seen, and so forth?
- Is each process step necessary?
- How much of your time is spent in rework?

Therefore, processing is dependent on

$$\text{Processing} = f(\text{WIP, old machine malfunction, unneeded process steps, WIP location, and so on})$$

Excess Inventory

Excess inventory is called storing excess products with no orders in the warehouse and having excess WIP. This will impede and tie up the cash flow. In fact, it may end up creating a negative cash flow. Here are some inventory value-added questions:

- Can you give a good reason for the extra inventory you have on hand?
- What about extra WIP?
- Is WIP pileup in certain areas unbalancing the line?
- What can you do about this?

Similarly, inventory is

$$\text{Inventory} = f(\text{cash flow, order, production, floor space, etc.})$$

Wasted Motions/Unutilized Talent

Movements that may cause injury in the manufacturing environment will result in process delay (review Section 5.2). Other unutilized talents are employee lost time, unused skills, employee ideas, and recommendations in simplifying the process. Here are some questions that will improve the process:

- How much walking do you do to complete an operation?
- Do you have to reach, bend, twist, or otherwise be uncomfortable to do processing or machine maintenance?
- How much time do you spend looking for things that get misplaced?

Wasted motions are dependent on the following factors:

Wasted motion $= f$ (injury, operator experience, mismatched operator talent, inefficient assignment to an experienced operator, lost time, etc.)

Errors and Defects

Defects will add additional rework, inspection (both expensive and time consuming), design changes, process changes, and machine downtime to analyze problems. In the plastics industry errors and defects include mold qualification time, engineering time, excess mold fabrication time and cost, and more production replacements. Errors in paperwork and engineering design will require additional time. The original cost must be absorbed, and unnecessary rework or replacement costs need to be captured. Here are some defect-related questions:

- What are the causes of poor products?
- Are processing mistakes occurring too frequently?
- Are you getting poor materials?
- Are machines malfunctioning?
- How much scrap and rework is avoidable?
- How many issues are repeat problems?

In the DOE model, defects are a function of the following:

Defects $= f$ (rework, inspection, process changes, design changes, scrap, paperwork, etc.)

So what do the seven waste types mean to an organization? They mean enormous work processes, increased or excess cycle time, and an increase in the waste products.

Consequently, the objective is to remove all the waste from the processing system and continue to keep it out. To ensure that this has been done efficiently and productively, the steps in Section 6.1.3 should result in the following:

1. Reduction of time
2. Reduction of space
3. Improvement in customer satisfaction, quality, profitability, and sales
4. Less machines, energy, and manpower being utilized in the process

6.1.3 THE FIVE LEAN THEMES AND STEPS

Value

This is value as defined by the customer. Understanding customer requirements and meeting them are the largest portion of the process needed to manufacture a product.

Value Stream

The value stream is the course of activities that is required to design, process,
build, assemble, and deliver to the customer (also see Section 5.2) from concept
through sales. The emphasis on value stream will improve return-on-investment
(ROI) values.

Value Stream Management
Value stream management is a technique of visually mapping the design of
    a product's manufacturing path from manufacturing warehouse to
    customer's door. Basically, the value stream is all the functions and
    actions needed to bring a product through the main essentials of the flow
    to the final product.

Value Stream Mapping
Value stream mapping will identify staff, information, and materials. It will
    also distinguish between value and nonvalue-added actions to improve
    value-added activities and reduce nonvalue-added actions. These are
    activities that external customers are willing to pay for. Value stream
    mapping is a visual flowchart that tracks materials, activities, and
    information required for the project. It is used to chart the existing and
    future process with a focus on value-added and nonvalue-added time.
 • *Value-added activities*: Activities in the process by which raw material or
    information is being transformed or shaped to meet customer
    requirements. Since product production or delivery lead time is critical in
    the project, one needs to be well organized and time coordinated.
    Continuously, we must keep the project updated and be cognizant of lean
    opportunities in the process.
 • *Nonvalue-added activities*: Activities or actions in the process that take
    service, resources, procedures, time, and space, but do not add to the

value of the product itself or do not improve upon external customer requirements. Consequently, nonvalue-added activities add additional work, inspection, cost, and so on. In other words, activities with little or no results are of no benefit to the customer.

Steps to Value-Added Analysis
1. Draw and complete a process flowchart for the project.
2. Distinguish all the job functions that add value to customer requirements, such as lower pricing, less defects, on-time delivery, and faster shipping.
3. Identify the nonvalue-added activities that do not add any value to the product—that is, inspection, counting, moving, reworking, or manual assembly in place of automation.
4. Select the activities that are important to be continued and actions that should be excluded or discontinued.

Since Lean means speed and velocity, which adds value to revenue and growth as well, one may ask how fast is quick enough to meet the optimum process. Basically the intensity of speed depends on the customer's timeline and requirements of how fast the process speed should be from supplier to the customer's door. This means the speed will meet the requirement without any nonvalue-added time.

Flow

Flow is the smooth progression of products or services. In smooth production flow, all the equipment and processing layout is planned in an orderly way so that work can flow without any distraction or interruption. This is one of the basic requirements of just-in-time (JIT) principles.

Pull Planning

Adopting a pull system requires determining the value, increasing the value of the value stream, and optimizing the flow value. In a pull design the system will pull the product through the process instead of introducing the raw material into the process. This is another rule of the just-in-time manufacturing system. This system is designed to reach the best possible quality, on-time delivery, lower cost of products or services to meet customer needs by the application of optimum flow, cycle time (Takt time), pull, and standardization concepts.

Perfection

This is the process by which activities create value and where waste is removed. In any industry, there are unlimited opportunities to improve the quality as well as a company's earnings. By exploring these opportunities and improving the quality and bottom line using elimination of waste, we can achieve the ultimate value at the lowest cost. On the other hand, perfection may never be achieved, but striving for it is worth the effort.

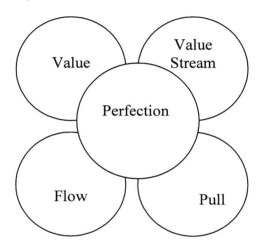

Management Support

A successful process improvement of any project requires management's approval and its strong support from the beginning of the project.

One way is to draw management's attention to the project's impact on the bottom line. The higher the return on investment (ROI), the stronger the management supports should be. To speed up the project and meet the deadline, these factors must be considered:

1. Pick a project with a higher ROI to put on top of the priority list.
2. Make sure all the members get involved with the project.
3. Members need to feel that they are part of this project. They need to know and feel that their expertise is needed.
4. Be supportive of other members' projects.
5. Update members and ask them to e-mail the results as soon as they are available.

6.2 THE ELEMENTS OF LEAN PERFORMANCE MEASUREMENTS

The measurements of all the factors that cause poor quality, waste, and defects are extremely important in the Lean *Six Sigma* processes. One question is "Why should we measure these factors?" We measure them for the following reasons:

1. What gets measured gets done. This is due to the fact that if we can't measure what we know in the form of mathematics, we really don't know much about it. And, usually, we do not question or comment on what we do not know much about. Nothing can be controlled if there is not enough information in the form of mathematical values about it. Therefore, we should know how to convert a process into numbers.

2. Process results should be recorded on a scorecard for performance reporting (see Appendixes).

3. Feedback loop measurement is needed for process improvement.

4. Management supports is required to achieve objectives.

5. Primary industrial measurement should include pressure, temperature, flow, design dimension, raw material cost/quality, inspection, lead time, testing time, and more.

From the beginning of the process, Lean measurement focuses on a vital few factors for higher impact, instead of many factors that produce diluted and time-consuming improvements:

- Measurement should be related to variables in achieving higher success, mission, vision, and values (see Section 6.2.1)—how organization performs now and in the future.
- Measurement should be applied to all sectors of organizations from the top to the bottom, as follows:
  — Corporate goals
  — Department goals
  — Individual goals
- Measurements should change as strategy changes.

Additional measurements for excellent performance include the strategic measurement model.

6.2.1 STRATEGIC MEASUREMENT MODEL

Mission, Vision, and Values
Any organization, in order to compete, succeed, and maybe even survive, must change or adopt new business strategies or corporate culture as time and the global economy changes. Mission, vision, and values are some of the strategic measurements companies should employ to examine their past, current, and future.

- What the organization is
  - Analyzing the organizational past, current growth, future growth, and success.

- The future goals of the organization
  - Achieving the future goals and dreams by using the current assets and resources already in the company and new planning as well as direction.
- What the organization stands for
  - To see and understand the company's value in the past, present, and even in the future.

At Hunter Industries, an irrigation company, the strategic initiatives to increase the company's value and profitability are stated in four steps:

1. Sales and revenue growth
   a. Grow sales of Hunter's product mix.
   b. Build specific support.
   c. Recognize and manage large customer growth.
   d. Develop international sales growth.
2. Products and services
   a. Research, develop, and launch new products and services.
   b. Design for maximum margin.
   c. Identify the unmet needs.
   d. Ensure quality and continuous process improvement.
3. Profitability
   a. Optimize manufacturing costs.
   b. Price and discount appropriately and profitably.
   c. Develop an improved internal organization and infrastructure.
4. People
   a. Hire quality people.
   b. Grow key talent.
   c. Retain those who are performing.
   d. Manage the performance of people.

With the preceding model, Hunter Industries has set the standard for the professional irrigation market. The company is regarded as the leader in what it does, with a reputation for valuable products, services, and people.

Key Success Factors and Business Fundamentals
These concentrate on what the organization must focus on to beat the competition and achieve its vision.

Performance Metrics
- Establish a balanced scorecard (see Appendix for a sample scorecard) that evaluates the basic changes in the organization. This helps in the creation of the economic future for the company. It is also an important statement to company leaders, due to the fact that normally

management benefits are bonded to financial objectives and objectives other than financial.
- Past-present-future

Most measurements and analyses are carried out on past performance. They also need to focus on the current and future projection performance.

Goals/Objectives
- The desired annual and long-term levels for each metric.

Strategies Planning
- Activities should be implemented to achieve the goals. This is short-term growth, which is normally based on one to three years and long-term growth that is usually based on five to ten years. Of course, this also could change yearly depending on the global economy growth and direction.

6.2.2 KEY ELEMENTS THAT MAKE A PRODUCT SUCCESSFUL IN THE MARKETPLACE

These elements are price, conformance, delivery speed, demand increases, product range, design, functionality performance, brand name, technical/customer support, after-sale support, and reliability that emphasize consistent quality and a uniform product over a specified period of time (product lifecycle). This is a time-oriented quality (Section 3.5 describes this in detail). In other words, product success is a function of the above variables such that Product success $= f$ (price, availability, on-time delivery, design, technical/customer support, and more).

The business performance measures consist of revenues, profitability (sales, cash flow, inventory), return on investment (ROI), return on assets (ROA), sales, market share, corporate growth, product cost, customer satisfaction (e.g., order fulfillment, on-time delivery, quality/price, build-upon schedule), cycle time, Takt time (e.g., the time that it takes a machine to produce a part or product), product yield, cost of quality, warranty costs, supplier performance, employee satisfaction, and employee safety.

6.3 COMPETITIVE PRODUCT BENCHMARKING CONCEPTS

Benchmarking is the process of identifying, understanding, and adopting outstanding practices from organizations anywhere in the world to help your organization improve its performance. Furthermore, it is a comparison measurement methodology to use as a reference point against the internal or external product concerning efficiency, quality, productivity, process, cost, and profitability. Most often comparison is used against a product or process made by external world class quality organizations. Many ongoing benchmarking methods are being used as evaluation techniques in different sectors of companies to improve and capture additional market share from the competition.

The following steps illustrate the benchmarking process:

1. Identify what to benchmark.
2. Form benchmarking teams.
3. Analyze the current process.
4. Establish benchmarking partners.
5. Analyze differences.
6. Search and identify best practices.
7. Initiate improvement actions.
8. Assess improvement; rebenchmark.

What should you benchmark? Here are Xerox's ten questions:

1. What is the most critical factor to my function's success?
2. What factors are causing the most trouble?
3. What products or services are provided to customers?
4. What factors account for customer satisfaction?
5. What specific problems have been identified in the organization?
6. What are the competitive pressures being felt in the organization?
7. What are the major costs in the organization?
8. What functions represent the highest percentage of cost?
9. What functions have the greatest room for improvement?
10. What functions have the greatest potential for market differentiation?

The benchmark process is shown in Figure 6.2.

Methods of Benchmarking

Benchmarking does not have to be from a similar company with the same product line (i.e., automobile companies A and B, or TV companies B and C). It could be any type of organization. The systems and techniques of manufacturing, production, delivery, lead time, and inventory are the key to success. The methodology makes the difference. The methods of improvement and functions can be transferred into any organization system. GE used the already developed *Six Sigma* models by Motorola and transferred its functions into the GE production line. Some of the methods are as follows:

1. Direct (customized) benchmarking can be done by survey or site-visit.
2. Indirect (general) benchmarking can be achieved by using online benchmarking companies or industry groups.

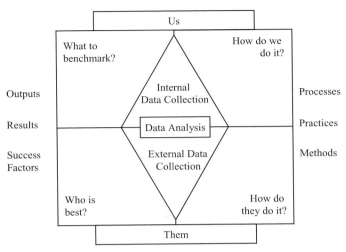

Figure 6.2   Benchmark process flowchart.

Figure 6.3 shows the graphical representation of Lean *Six Sigma* improvement steps.

6.4 INTEGRATION OF KAIZEN, LEAN, AND *SIX SIGMA*

6.4.1 *Six Sigma*, Lean, and Kaizen Principles

Six Sigma, Lean, and Kaizen, each in its own way, are the major strategic problem-solving techniques in the industrial world. The integration of all three quality methods will bring the most powerful tool to eliminate waste and improve productivity and profitability, as shown in Figure 6.4. The *Six Sigma* continuous improvement toward customer satisfaction and bottom line, Lean speed, and reduction in unnecessary wait time were discussed in Chapters 4 and 5. Here the Kaizen action-based principle will be reviewed (see the details in Chapter 8). The term *Kaizen* is the Japanese word for "continuous improvement." Basically, any action toward the improvement of processes is called Kaizen.

The Kaizen strategies are based on several rules that may differ from one organization to another, depending on their application. They are composed of the following action-based processes:

1. Keep a positive attitude.
2. Be open-minded.

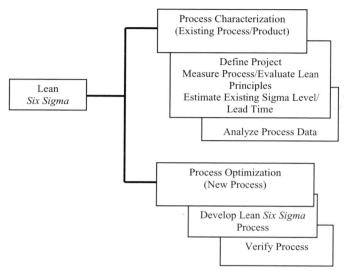

Figure 6.3 Representation of Lean *Six Sigma* DMADV (define/measure [diagnosis], analyze, develop, verify) levels.

3. Get involved in teamwork and challenges.
4. No excuses—look for solutions.
5. Ask the five *whys* rule (see Section 8.4.1.2 for more details).
6. Apply ideas immediately with available resources and do not wait for perfection.
7. Treat and implement all the team members' opinions equally. Ignore member rank.
8. Apply 5S improvements rules:
   - S_I = Sort
     - Sort out what you do not need. For instance, review and analyze filing cabinets, documents, chairs, computers, phones, equipment, and so on. Eliminate anything that is obsolete. Keep what you need, and throw away unneeded material.
   - S_{II} = Set in order
     - Organize (what, where, and how).
       Organize the leftovers from the sorting process, using efficient and effective storage techniques. Here are some examples:
       a. One strategy is painting the floor for process arrangement.
       b. Just like filing cabinets, arrange items in alphabetical or by application order. Make them easy to locate and use.
       c. Items needed most or frequently used should be stored nearby.

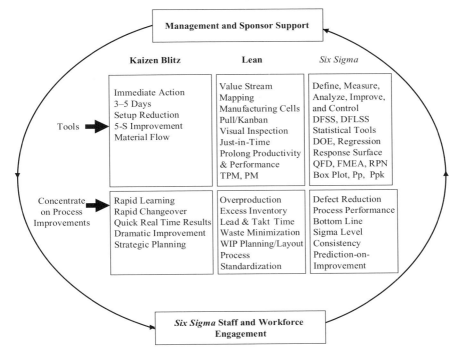

Figure 6.4   Integration of Kaizen Blitz, Lean, and *Six Sigma* process flowchart.

- S_{III} = Shine
- Once sorting (S_I) and organizing (S_{II}) have been completed, the next step is cleaning the area—for example, bookshelves, equipment, storage area, conference room, office rooms, under the table, and so on.
  - What to clean
    Make a checklist of what to clean.
  - How to clean
    Cleaning procedure also needs to be stated.
  - Who will clean?
    Personnel and responsibilities should be determined. Regularly, cleaning areas need to be checked.
  - How much cleaning is necessary?
    Make cleaning a daily habit.
- S_{IV} = Standardize
  - It has the improved and possibly the best process, and its steps have been standardized (or documented to be standardized).

Now, once S_I through S_{III} have been completed, to keep everything clean, we must standardize by making it part of the daily routine job.

- S_V = Sustain
  - Maintain the process.
  Control and maintain the standardized 3S (S_{IV}) and incorporate with corporate culture.

9. Apply just-in-time (JIT) concepts.

JIT by definition is making specified required parts available at a specified required place and at a specified required time. For instance, a manufacturing company's assembly department receives the right amount of inventory at the right time without stocking the parts.

JIT, which originated at Toyota, is how material or parts should be moved in order to be ready for the next step in production (assembly). Its model uses demand, pull, Takt time, and Kanban. It starts with reducing inventory and using a two-bin Kanban technique. The outcome is less inventory and short Takt time for production. Key elements of JIT are flow, pull, Takt time, and standard in process inventory. Thus, JIT methodology is dependent on balancing the production by reducing inventory so that just the needed amount of product is ordered by the customer, and it is delivered on time without any delay.

6.4.2 PROLONG PRODUCTION PERFORMANCE (PPP)

Both today and in the future, as the population increases, automation also increases and becomes an integral part of our lives. Process by human becomes process by machine (or process by robot). Automated process not only exists in manufacturing and production assembly, but it also has expanded to health care, transportation, quality measurement, even education by computer (workstation or online lectures), laboratories, and communication (TV, phone, and many more).

The goal is no machine downtime, zero defects, maximum machine lifecycle and throughput, customer satisfaction, and the bottom line. Basically, the goal is to prolong production performance until it reaches a point that the machine requires complete replacement due to wear and tear or technology change, if justified. The following are some of the highlights and examples of tools to achieve this goal:

1. In the case of injection molding machines, manufacturers should obtain machines that are designed to function with the designated material properties (physical/mechanical), or they need to select product materials that minimize

interference. The process is then required to maintain the actual machine productivity rate and reduce the cycle time in order to utilize machine design features as well as productivity goals.

2. Analyze and improve the root cause of interference of downtime if it is molding issues, molding machine issues, operators, setups, material, tool/part design, or uncontrollable issues.

3. Train operators in product inspection and basic maintenance issues and troubleshooting techniques. Hold them accountable for their job function and responsibilities.

4. Adjust break time schedules with relief staff to enable the continuation of production; otherwise, each downtime may add other defects before the machine stabilizes itself.

5. Adopt total productivity maintenance (TPM) and preventive maintenance (PM) techniques in achieving optimum productivity goals in addition to the preceding steps.

Total Productivity Maintenance (TPM)

TPM was originally developed in Japan in the 1970s. This technique is based on the promotion of a PM concept that began in the United States in the 1950s to utilize tools and achieve the following benefits:

TPM prolongs (or maximizes) the machine/tool lifecycle and productivity by promoting PM through leadership motivation. This involves all of the employees in every department of an organization (e.g., maintenance, engineering, operators, supervisors, etc). These are some highlights of TPM principles:

- Engage top to bottom shop management in the support of the TPM program.
- Train, educate (improve skills), and promote TPM within the company.
- Establish regulations and policies.
- Develop a process map and plan TPM objectives.
- Evaluate and improve the effectiveness of equipment.
- Plan and implement a maintenance department program.
- Apply TPM and continuously improve program advancement.
- Adopt any new maintenance technologies that are applicable and can be justified.

When the leadership team understands *Six Sigma*, Lean, and Kaizen, they can provide clearer direction on what must be done to improve profitability and competitiveness.

The *Six Sigma*/DFSS, Lean, and Kaizen thinking will dramatically boost customer satisfaction and profitability, while sharply reducing waste, defects, errors, accidents, and time to market. To improve organizational performance, the new economic strategies require industries to implement *Six Sigma*, Lean, and Kaizen processes. The Kaizen further expands to the Kaizen Blitz theory.

The Kaizen Blitz Events

Kaizen Blitz (a German term meaning "lightning flash") is a methodology that permits you to identify the opportunities for continuous improvement in a short period of time, such as days or hours. It is an action-based strategy to apply Lean, JIT, quick result, and time-to-market. In business, results should transform to increase income. But a well-established company focuses on a strategy with much more financial improvement than projected (e.g., if a student wants to achieve a "B" grade on a course, he must put "A" grade effort to solidify the "B" letter. An effort for "B" may not get the student a "B"). These are the Kaizen Blitz objectives:

1. Team-based workplace improvement. Select and organize team members, particularly those who support Kaizen Blitz. Inform them of the steps and possible outcome.
2. Sense of urgency.
3. Immediate resource availability.
4. Creativity before capital.
5. Quick and crude (inelegant) versus slow and elegant.
6. Immediate results.
7. Follow-up. Standardize the improvements (or profits), and apply them to ongoing processes.

These are the Kaizen Blitz processes:

1. Measure the existing performance.
2. Examine the process.
3. Develop new work combinations.
4. Redesign flowchart layouts.
5. Apply implementation.
6. Measure results and communicate to participants.
7. Identify longer range executions.

Just like *Six Sigma*, the Kaizen Blitz team must be cross-functional, self-directed, and consist of process-improving individuals. The following are the Kaizen Blitz goals:

1. Obtain quick results.
2. Rearrange the process.
3. Employ better, not necessarily the best, solutions.
4. Faster learning by doing.
5. Quick team development.

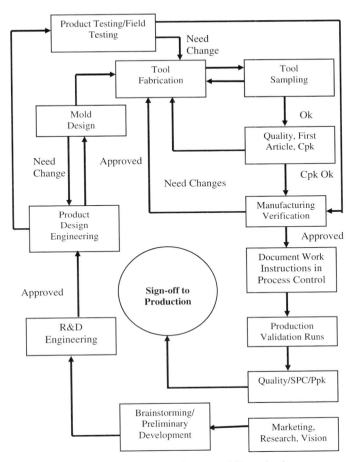

Figure 6.5   Process map for automobile production.

6.4.3 A LEAN CONCEPT IN REDUCTION OF LEAD TIME

An automotive manufacturing company is looking for techniques to reduce lead time between processes in order to meet the new version timelines by October 1 of each year. The processes include design, fabrication, sampling, quality, manufacturing, and engineering (engineering changes and manufacturing changes). A sample process map is shown in Figure 6.5. The steps in Figure 6.5 will occur if (1) the final deadline is enforced, (2) the timeline for each step is established, and (3) the entire loop (cycle) is achieved and completed in no more than three trials.

One case study analyzing the process showed that time traps occurred during engineering testing and evaluation. Basically, engineering change order (ECO) and manufacturing change order (MCO) sign-off took the longest. The longer the ECO/MCO sign-off period, the more loss of profit and market share.

Here are some of the lead time observed and suggested improvement processes:

1. Thorough design, analysis with extensive study of current technologies, and previous design implication.

2. Implementation of all resources before tool fabrication to achieve optimum and ultimate tool design within the budget.

3. Utilize preselected and proven materials (i.e., steel, plastics). If new material is selected, then the following steps will improve the lead time, in addition to all the testing.

   a. To reduce lead time, the properties of the material should be studied for dynamics of fluid flow, physical, mechanical, chemical, biological, microbial, corrosion, global economy effect on cost, and availability due to consumption, quality, failures, and legal implications.

   b. Avoid trial-and-error testing, as well as material selection. Pick the top few materials that are based on research and consultation with experts on those materials (do a thorough material study).

4. Creation of a well-determined, confident, and united team with a "can do" attitude for the life of the project.

5. Creation of a goal-oriented team to meet the timeline. Sometimes an award may expedite both the process and the team member's attitude. For instance, in one case, a major *Six Sigma* company was not moving well after two years of a *Six Sigma* trial program. The CEO stepped in and said 40% of all management/employee year end rewards will come from the *Six Sigma* program. This was the major boost in the project and the company started to earn tangible rewards.

6.5 LEAN/KAIZEN *SIX SIGMA* INFRASTRUCTURE EVOLUTION TOOLS AND HIGHLIGHTS IN SUMMARY

6.5.1 CORPORATE COMMITMENT

The CEO and the highest management of a corporation should be committed and engaged with full support of the Lean *Six Sigma* initiative throughout the program. The cultural and infrastructure changes must be accepted by senior and top management for the future vision of the company, which include Lean operation, customer satisfaction, and bottom-line awareness. Furthermore, all the managers should be trained and educated on the program.

The success of Lean *Six Sigma* is a function of CEO passion and commitment. As mentioned in Chapter 1, goals (i.e., financial, market share) must be set for two to five or six years and achieved in accordance with the timeline (target date). For instance, goals should be like the following:

- For the current fiscal year, each Black Belt project must return on investment $300,000–$500,000 (i.e., year 2007 standard).
- Number of field returns on goods should be reduced by 20 to 30%.
- Corporate scrap rate must be reduced to .1 percent by the next two years.
- Inventory level must be reduced 50%.

The sigma level of entire projects each year must show improvement.

6.5.2 STEPS TO ACHIEVE THE *SIX SIGMA* GOALS

1. Define a project where improvement will have the highest impact (defect reduction), either in manufacturing or in a transactional area. Always form a mathematical function related to the problem—for example, $y = f(x)$, where y is output and x is input.
2. Measure the defect in the form of numbers and collect data, transactional (i.e., lead time) or manufacturing (i.e., dimensional, mechanical, quality), for the existing process in the operation.
   - Lead time, is a nonproduction/manufacturing defect. A longer lead time has an enormous impact on profitability and financial goals to such an extent that its elimination could add huge profits to the bottom line.
3. Analyze the collected data using Lean *Six Sigma* techniques for Lean, process capability, cause and effect, nonvalue-added process, and so on.
4. Develop/improve the analyzed data by applying Kaizen, Lean and *Six Sigma*, JIT, TPM, PPP, 5S, value stream, Poka-Yoke, laying out financial objectives, and the aim of the project. Always set a deadline.
5. Verify/control or maintain the improved process by establishing control charts/Lean matrix and continue to improve the speed, lead time, process capability, sigma level, and reduction of defects.

Some of the *Six Sigma* projects may look linear, but they will be mostly nonlinear. The challenge is how to convert the nonlinearity to linearity. Without any systematic commitment, direction, and knowledge of the possible tools, it is impossible. If there is no commitment from top management, it would be better not to start the program because it will fail.

The preceding steps with tools and techniques for Green and Black Belt are shown in Figures 6.6 and 6.7.

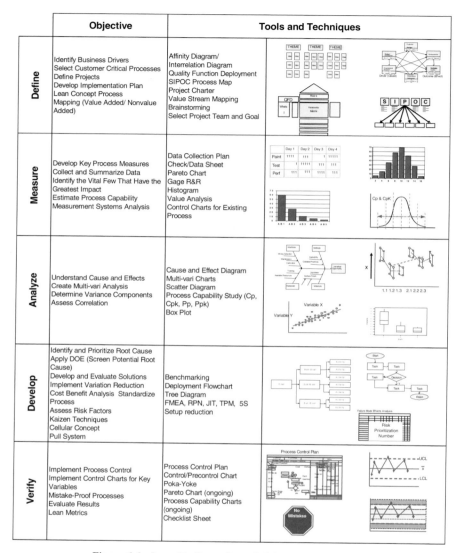

	Objective	Tools and Techniques
Define	Identify Business Drivers Select Customer Critical Processes Define Projects Develop Implementation Plan Lean Concept Process Mapping (Value Added/ Nonvalue Added)	Affinity Diagram/ Interrelation Diagram Quality Function Deployment SIPOC Process Map Project Charter Value Stream Mapping Brainstorming Select Project Team and Goal
Measure	Develop Key Process Measures Collect and Summarize Data Identify the Vital Few That Have the Greatest Impact Estimate Process Capability Measurement Systems Analysis	Data Collection Plan Check/Data Sheet Pareto Chart Gage R&R Histogram Value Analysis Control Charts for Existing Process
Analyze	Understand Cause and Effects Create Multi-vari Analysis Determine Variance Components Assess Correlation	Cause and Effect Diagram Multi-vari Charts Scatter Diagram Process Capability Study (Cp, Cpk, Pp, Ppk) Box Plot
Develop	Identify and Prioritize Root Cause Apply DOE (Screen Potential Root Cause) Develop and Evaluate Solutions Implement Variation Reduction Cost Benefit Analysis Standardize Process Assess Risk Factors Kaizen Techniques Cellular Concept Pull System	Benchmarking Deployment Flowchart Tree Diagram FMEA, RPN, JIT, TPM, 5S Setup reduction
Verify	Implement Process Control Implement Control Charts for Key Variables Mistake-Proof Processes Evaluate Results Lean Metrics	Process Control Plan Control/Precontrol Chart Poka-Yoke Pareto Chart (ongoing) Process Capability Charts (ongoing) Checklist Sheet

Figure 6.6   Lean *Six Sigma* Green Belt improvement model.

	Objective	Tools and Techniques
Define	Identify Business Drivers Select Customer Critical Processes Define Projects Develop Implementation Plan Lean Concept Process Mapping (Value Added/ Nonvalue Added)	Affinity Diagram/ Interrelation Diagram Quality Function Deployment SIPOC Process Map Project Charter Value Stream Mapping Brainstorming Select Project Team and Goal	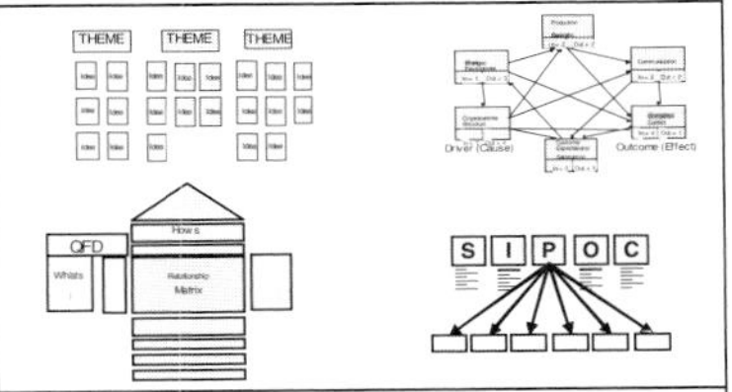
Measure	Develop Key Process Measures Collect and Summarize Data Identify the Vital Few That Have the Greatest Impact Estimate Process Capability Measurement Systems Analysis	Data Collection Plan Check/Data Sheet Pareto Chart Gage R&R Histogram Value Analysis Control Charts for Existing Process	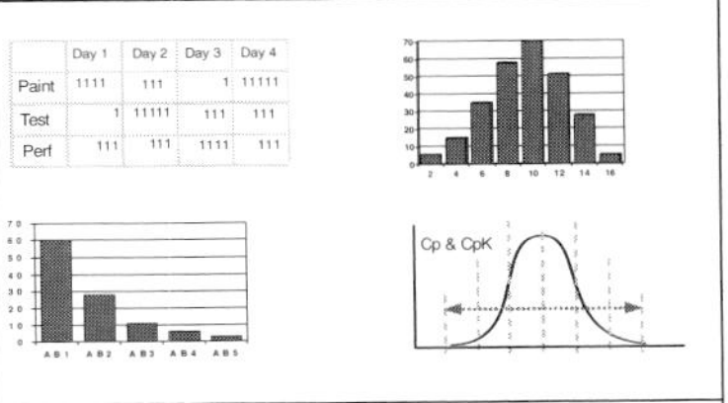
Analyze	Understand Cause and Effects Create Multi-vari Analysis Determine Variance Components Assess Correlation	Cause and Effect Diagram Multi-vari Charts Scatter Diagram Process Capability Study (Cp, Cpk, Pp, Ppk) Box Plot	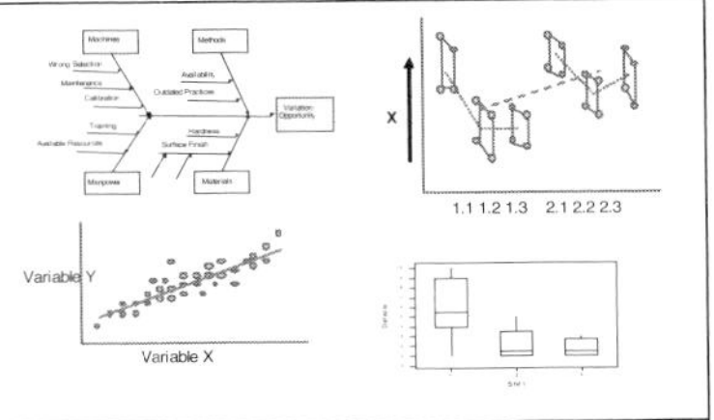
Develop	Identify and Prioritize Root Cause Apply DOE (Screen Potential Root Cause) Develop and Evaluate Solutions Implement Variation Reduction Cost Benefit Analysis Standardize Process Assess Risk Factors Kaizen Techniques Cellular Concept Pull System	Benchmarking Design of Experiments Response Surface Methodology Deployment Flowchart Tree Diagram FMEA, RPN, JIT, TPM, 5S Setup reduction	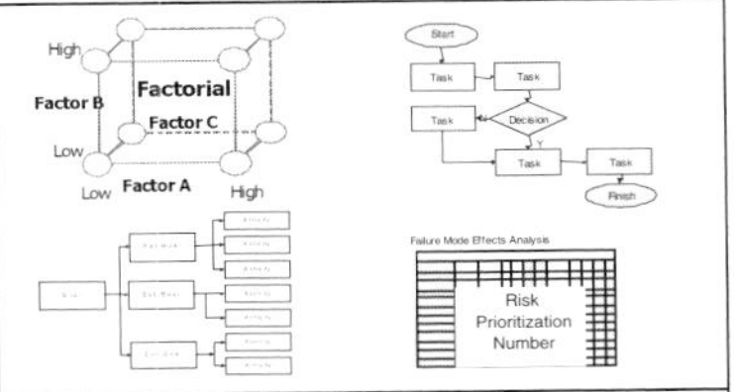
Verify	Implement Process Control Implement Control Charts for Key Variables Mistake-Proof Processes Evaluate Results Lean Metrics	Process Control Plan Control/Precontrol Chart Poka-Yoke Pareto Chart (ongoing) Process Capability Charts (ongoing) Checklist Sheet	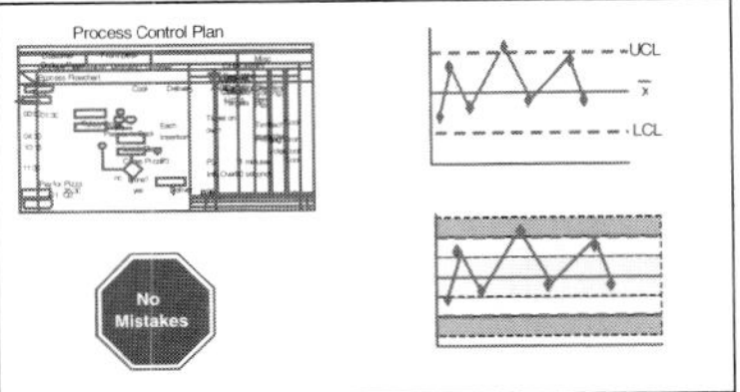

Figure 6.7 Lean *Six Sigma* Black Belt improvement model.

6.6 MATHEMATICAL MODELING OF LEAN *SIX SIGMA* RELATIONS

Equation 3.3 defines *output* as a function of input. This means that output changes are dependent on input changes (or independent variable).

$$\text{Output} = f(\text{Input})$$

With that in mind, let's define the basic fundamentals of customer focus and follow the success model of any business, such as the following:

$$\text{Business success} = f\,(\text{customer satisfaction, bottom line})$$

This defines business success as a function of (or dependent on) customer satisfaction and bottom line, as shown:

Customer satisfaction and bottom line $= f$ (quality, price, availability, and shipping)

Furthermore,

Quality, price, availability, and shipping $= f$ (process performance, C_{PK}, P_{PK})

Likewise process performance is as follows:

Process performance $= f$ (process variation (as cycle time, lead time))

and, moreover, process variation depends on

Process variation $= f$ (statistical and technical knowledge of employee)

and finally:

$$\text{Employee knowledge} = f\,(\text{training, education})$$

This concept applies to world class performance (WCP). The response is a function of Lean *Six Sigma* previously given in the first and second rounds of measurement in Equations 6.1 and 6.2.

The design of experiment (DOE) can be applied to all the steps of Lean *Six Sigma* and its principles. Using the response function, one may even expand DOE further to regression analysis or future prediction curve. Therefore:

World class performance (WCP) $= f$ (Lean *Six Sigma* (measurement variables of lead time, design, and process))

Similarly, process performance (y_{pp}) is a function of design, material, tooling, and machine or process condition.

Process performance (PP) $= y_{pp} = f$ (product design, material, tooling, and process condition)

In terms of mathematical symbols, the response function can be written as

$$y_{pp} = f(x_1, x_2, x_3, \ldots, x_{n-2}, x_{n-1}, x_n) \tag{6.3}$$

Here, y_{pp} and y_{wcp} (refer to Equation 6.1) are the predicted responses for the given Lean *Six Sigma* independent variables (or causes of variation) $x_1, x_2, x_3, \ldots, x_{n-2}$, x_{n-1}, x_n, such as pressure, temperature, cycle time, material moisture level, machine downtime, signature approval wait time, rework, inspection time, inspection cost, and the list goes on and on. The polynomial representation of Equation 6.2 or 6.3 (process behavior) for one through three variables in quadratic and cubic models is as follows:

Linear model

$$y = b_0 + b_1 x_1 + b_2 x_2 + b_3 x_3 \tag{6.4}$$

Likewise, quadratic two-variable polynomial:

$$y = b_0 + b_1 x_1 + b_2 x_2 + b_{11} x_1^2 + b_{22} x_2^2 + b_{12} x_1 x_2 \tag{6.5}$$

and three-variable polynomial model:

$$
\begin{aligned}
y = {}& b_0 + b_1 x_1 + b_2 x_2 + b_3 x_3 && \text{Linear terms} \\
& + b_{11} x_1^2 + b_{22} x_2^2 + b_{33} x_3^2 && \text{Quadratic effects} \\
& + b_{12} x_1 x_2 + b_{13} x_1 x_3 + + b_{23} x_2 x_3 && \text{Binary interaction effects} \\
& + b_{123} x_1 x_2 x_3 + b_{112} x_1^2 x_2 + b_{113} x_1^2 x_3 \\
& + b_{122} x_1 x_2^2 + b_{133} x_1 x_3^2 + b_{223} x_2^2 x_3 \\
& + b_{233} x_2 x_3^2 + b_{111} x_1^3 + b_{222} x_2^3 + b_{333} x_3^3 && \text{Cubic interaction effects}
\end{aligned}
\tag{6.6}
$$

Equations 6.5 and 6.6 are response surface models for multiple variables in Lean *Six Sigma*, where constant b_0 is the response intercept and $b_1, \ldots, b_{333}$ in Equation 6.6 are called regression coefficients of variables (these are unknown in the model), which can be estimated using the experimental data (without experimental data b's cannot be determined). Or any statistical software will calculate the coefficients, such as Minitab. These equations also include the interaction between the causes of variation, which is not possible to measure in a linear system.

6.6.1 LEAN *SIX SIGMA* EXPERIMENTAL DESIGN

An experimental design is a method of systematically laying out a detailed experiment (or plan) before carrying out the project. It is also a technique in reducing the cost of research in a timely manner and process understanding by avoiding trial-and-error procedures. Unlike trial and error, data obtained from DOE can be converted to a mathematical model, which could be used in a computer for process optimization and simulation. The theory of DOE normally starts with the idea of the process models illustrated in Figure 6.1. However, most empirical models (or equations) of experimental data are either linear (first-degree polynomial) or quadratic or nonlinear (second-degree polynomial) in form and often include third degree polynomials (cubic), as shown in Equation 6.6.

There are numerous designs of experiment (DOE) methods available in books on DOE. One of the most popular is Central Composite Design (CCD), particularly for second-degree polynomials. The CCD uses full factorial treatment. It is an accurate design with all the possible runs for specified variables. However, economically, for any treatment higher than three variables, it gets to be a more expensive process because full factorial uses all levels of each factor. Thus, other methods, such as fractional factorial, come into the picture, which use fewer runs than full factorial. The choice or type of the experiment depends on the financial availability for experimenting, as well as time and resources.

The success of DOE depends on the methodology of each design and factor selection. To expedite the factor selection, one may design a fishbone or cause-and-effect diagram, with people closest to the process. Or conduct a DOE with multiple factors and determine the most influential factors after the analysis. Then, using the factors that had the highest impact in the process, design a new experiment to find out the effects of each factor on the quality. However, any design should have as many unknown runs as possible. In any experiment, the higher the number of runs, the higher the accuracy of the experiment and of the prediction model.

EXAMPLE 6.1. Problem Statement: Minimizing variability.

A manufacturing quality control (QC) department has noticed variations in parts' diameters (possible shrinkage) after they are removed from the machine and cooled off for about four hours. This will lead to customer dissatisfaction and will affect the bottom line with an increase in scrap if the problem is not fixed. Thus, QC asked the process engineer to examine the case. A process engineer was assigned to improve the diameter (y) of a plastic part formed by an injection molding machine.

The first process engineer decided to assemble all the people who work closely with the process and brainstorm some of the factors that affect the part quality by establishing a fishbone or cause-and-effect diagram for injection molding. The key functions of the injection molding machine and fishbone (cause-and-effect) diagram are shown in Figures 6.8 and 6.9.

The process model for Example 6.1 is illustrated also in Figure 6.10.

Then, the process engineer decided to do a prescreen experiment to find the most influential factors with the following independent variables. Notice that +1 indicates the higher-level setting and −1 indicates the lower-level setting, as shown in Table 6.2.

An eight-run fractional factorial design (such as Taguchi 2-level 4 factors) is carried out with four replicates of each run using Tables 6.3 and 6.4.

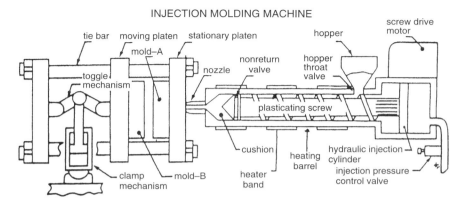

A Typical Screw Injection Molding Machine

Figure 6.8   Key functions of injection molding machine.

Table 6.2

Cause Factors (Independent Controllable Variables)

	Controllable variables name	Low level (−1)	High level (+1)
1	A = Plastic cooling time (x_1)	12 sec	16 sec
2	B = Plastic pack pressure (x_2)	10,000 psi	14,000 psi
3	C = Plastics melt temperature (x_3)	350°F	400°F
4	D = Mold surface temperature (x_4)	160°F	190°F

Table 6.3

Fractional Factorial Coded Design Matrix (Four Factors at two Levels)

	A	B	C	D (ABC) x_4 ($x_1x_2x_3$)	AB (CD) x_1x_2 (x_3x_4)	AC (BD) x_1x_3 (x_2x_4)	AD (BC) x_1x_4 (x_2x_3)
	x_1	x_2	x_3				
1	−1	−1	−1	−1	+1	+1	+1
2	−1	−1	+1	+1	+1	−1	−1
3	−1	+1	−1	+1	−1	+1	−1
4	−1	+1	+1	−1	−1	−1	+1
5	+1	−1	−1	+1	−1	−1	+1
6	+1	−1	+1	−1	−1	+1	−1
7	+1	+1	−1	−1	+1	−1	−1
8	+1	+1	+1	+1	+1	+1	+1

Molding Process Optimization Variables

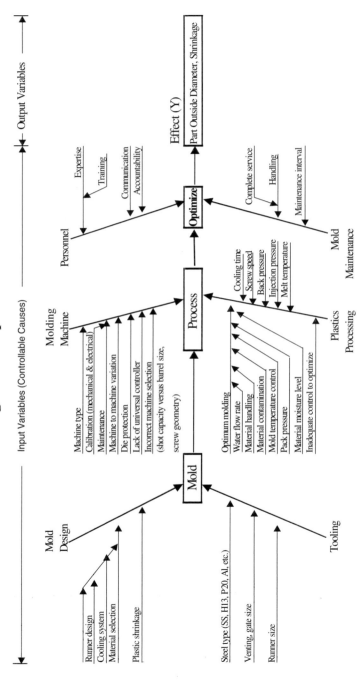

Figure 6.9 Root cause analysis of part outside diameter/shrinkage.

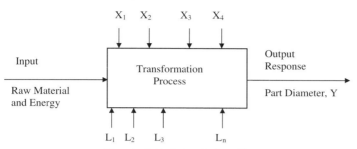

Figure 6.10   Process model.

Table 6.4

**Actual (Uncoded) Values of Experimental Matrix with Response
before Experimentation**

	x_1 (sec)	x_2 (psi)	x_3 (°F)	x_4 (°F)	y_1	y_2	y_3	y_4	y ave.	Std. dev.
1	12	10,000	350	160						
2	12	10,000	400	190						
3	12	14,000	350	190						
4	12	14,000	400	160						
5	16	10,000	350	190						
6	16	10,000	400	160						
7	16	14,000	350	160						
8	16	14,000	400	190						

Since the mold temperature profile changes require longer time to stabilize, all the experimental design runs were carried out randomly rather than sequentially, as shown in Table 6.5. This was done to reduce the process lead time. All the parameter settings other than factors in Table 6.2 were kept constant. The nominal diameter in this experiment was given as 1.665 (±0.002).

Table 6.6 converts Table 6.5 data in a sequential format (just like Table 6.4).

Now the experiment has been completed, and it is time to analyze the data as described in the analysis phase.

Analysis Phase

Phase (1): Table 6.6 shows that runs #3 and #5 are the closest to experiment nominal (1.665).

Design for Lean/Kaizen Six Sigma

Table 6.5

Actual Values of Experimental Matrix with Random Runs

Run	x_1 (sec)	x_2 (psi)	x_3 (°F)	x_4 (°F)	y_1 (in)	y_2 (in)	y_3 (in)	y_4 (in)	y ave. (in)	Std. dev. (s)
1	12	10,000	350	160	1.661	1.660	1.662	1.651	1.661	0.0051
4	12	14,000	400	160	1.665	1.669	1.666	1.668	1.667	0.0018
6	16	10,000	400	160	1.663	1.664	1.661	1.664	1.663	0.0014
7	16	14,000	350	160	1.668	1.669	1.666	1.669	1.668	0.0014
2	12	10,000	400	190	1.662	1.662	1.663	1.661	1.662	0.0008
3	12	14,000	350	190	1.668	1.665	1.667	1.664	1.666	0.0018
5	16	10,000	350	190	1.664	1.662	1.667	1.663	1.664	0.0022
8	16	14,000	400	190	1.668	1.666	1.668	1.670	1.668	0.0016
Overall Average									1.665	0.002

Table 6.6

Actual Values of Experimental Matrix with Random Runs

Run	x_1 (sec)	x_2 (psi)	x_3 (°F)	x_4 (°F)	y_1 (in)	y_2 (in)	y_3 (in)	y_4 (in)	y ave. (in)	Std. dev. (s)
1	12	10,000	350	160	1.661	1.660	1.662	1.651	1.661	0.0051
2	12	10,000	400	190	1.662	1.662	1.663	1.661	1.662	0.0008
3	12	14,000	350	190	1.668	1.665	1.667	1.664	1.666	0.0018
4	12	14,000	400	160	1.665	1.669	1.666	1.668	1.667	0.0018
5	16	10,000	350	190	1.664	1.662	1.667	1.663	1.664	0.0022
6	16	10,000	400	160	1.663	1.664	1.661	1.664	1.663	0.0014
7	16	14,000	350	160	1.668	1.669	1.666	1.669	1.668	0.0014
8	16	14,000	400	190	1.668	1.666	1.668	1.670	1.668	0.0016
Overall Average									1.665	0.002

Phase (2): Construct the average, effects, and half effects for the factors and interactions of responses and standard deviations using Table 6.6.

These are shown in Tables 6.7 and 6.8.

Phase (3): Create a plot of the response averages and standard deviation at the high and low values for each effect, which are shown in Figures 6.1 through 6.13.

Phase (4): Create a Pareto chart of the absolute value of each effect for response and standard deviations that are shown in Figures 6.14 and 6.15.

Phase (5): This phase will determine the important factors. To find the critical variables that shift the response average and the standard deviation of responses, one can look at a Pareto chart of the average and standard deviation effects in Figures 6.12 and 6.13. According to Figure 6.11 and the statistical background, the factors A and B and, to some extent, the interaction AC are the factors to be

Table 6.7

Response Effects of Factors/Interactions

Cause/interaction	y− Ave.	y+ Ave.	Δ Effect	Δ½ Effect
A = Plastic cooling time (x_1)	1.664	1.6665	0.0025	0.00125
B = Plastic pack pressure (x_2)	1.6625	1.6673	0.0048	0.0024
C = Plastics melt temperature (x_3)	1.6648	1.665	0.0002	0.0001
D = Mold surface temperature (x_4)	1.6648	1.665	0.0002	0.0001
AB (CD)	1.665	1.6648	−0.0002	−0.0001
AC (D)	1.6653	1.6645	−0.0008	−0.0004
AD (BC)	1.6648	1.665	0.0002	0.0001

Table 6.8

Standard Deviation Effects of Factors/Interaction

Cause/interaction	s− Ave.	s+ Ave.	Δ Effect	Δ½ Effect
A = Plastic cooling time (x_1)	0.00238	0.00165	−0.00073	−0.00037
B = Plastic pack pressure (x_2)	0.002375	0.00165	0.00073	−0.000363
C = Plastics melt temperature (x_3)	0.002625	0.00140	−0.00123	0.0001
D = Mold surface temperature (x_4)	0.002425	0.0016	−0.00083	0.0001
AB (CD)	0.0018	0.00223	0.00043	0.000215
AC (D)	0.00155	0.00248	0.00093	0.000465
AD (BC)	0.00135	0.00268	0.00133	0.000665

considered. But, since AC and D are aliased, it is difficult to determine whether the effect is because of AC or D. Further, due to the fact that A and B are stronger candidates than AC, A and B both will be considered for further study.

Phase (6): Develop a response prediction equation:

$$y = b_0 + b_1 x_1 + b_2 x_2 + b_{12} x_1 x_2$$

or

$$y = b_0 + b_1 A + b_2 B + b_{12} AB$$

$$y = \bar{y} + \left(\frac{\bar{A}_+ - \bar{A}_-}{2}\right) A + \left(\frac{\bar{B}_+ - \bar{B}_-}{2}\right) B + \left(\frac{\overline{AB}_+ - \overline{AB}_-}{2}\right) AB$$

$$y = \bar{y} + \left(\frac{\Delta}{2}\right) A + \left(\frac{\Delta}{2}\right) B + \left(\frac{\Delta}{2}\right) AB$$

Using half response effect values from Table 6.7, the prediction model looks like

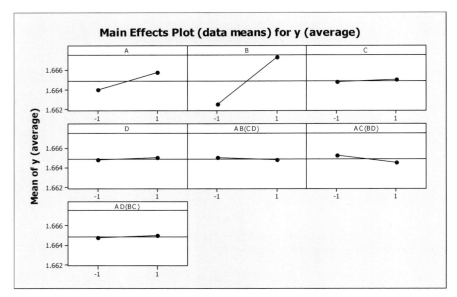

Figure 6.11   Plot of the response averages for each effect.

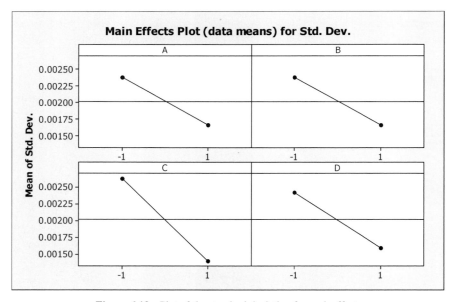

Figure 6.12   Plot of the standard deviation for each effect.

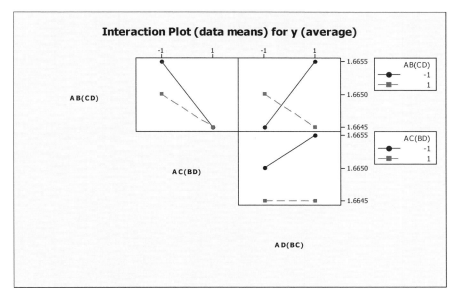

Figure 6.13   Plot of the response averages for each interaction.

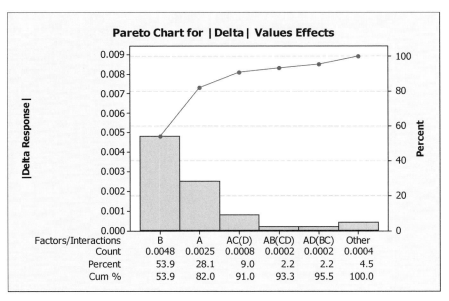

Figure 6.14   Absolute value of delta response effect Pareto chart.

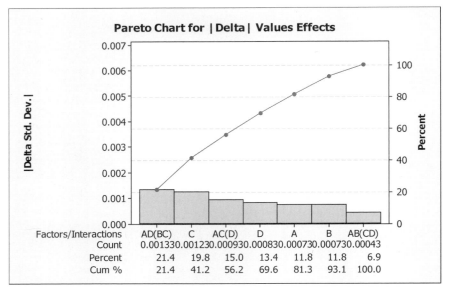

Figure 6.15   Absolute value of delta standard deviation effects Pareto chart.

$$y = 1.665 + 0.00125A + 0.0024B + (-0.0001)AB$$

To develop the standard deviation prediction model for orthogonal design, the same method of modeling for response y can be used by implementing information from Table 6.8 and the Pareto chart in Figure 6.15.

EXAMPLE 6.2.   Analysis of Example 6.1 indicates that the highest impacts in the process factors are plastics pack pressure x_1 or A and plastics cooling time x_2 or B. A two-level full factorial design using factors A and B will be developed to optimize the process and reduce the part diameter variation as follows.

The variables in the Figure 6.7 process model are reduced to x_1 and x_2, as shown in Figure 6.16. Table 6.9 represents the variables with high and low values. To keep the experiment consistent, all the high and low values will be the same as a screening experiment.

A very effective and efficient full factorial second-order (modeling) design, which was first introduced by Box-Wilson, is called central composite design (CCD) and will be applied here. A CCD is designed rotatable for α. The rotatability of α depends on the number of runs in the factorial segment (Equation 6.6) of CCD. Rotatability has advantages in the designing of response surface methodology (RSM is an optimization technique; see Chapter 10). The optimum

Table 6.9

Cause Factors (Independent Controllable Variables)

	Controllable variables' name	Low level (−1)	High level (+1)
1	A = Plastic cooling time (x_1)	12 sec	16 sec
2	B = Plastic pack pressure (x_2)	10,000 psi	14,000 psi

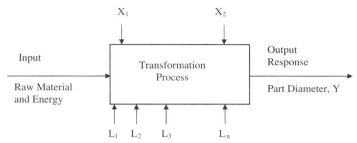

Figure 6.16 shows the Lean *Six Sigma* controllable variables; plastic pressure and temperature.

X_1 and X_2 feed into Transformation Process.

Input — Raw Material and Energy.

Output Response — Part Diameter, Y.

L_1 L_2 L_3 ... L_n

Uncontrollable factors in Lean *Six Sigma* process, such as machine loads and noise (electric), etc.

Figure 6.16   Process model.

point of any design is unknown in advance of experimental runs. The value of α can be calculated as shown in Equation 6.6.

$$\alpha = \left(L^k\right)^{\frac{1}{2}} \tag{6.7}$$

where L^k is the number of factorial trials. In the case of Example 6.2 the Equation 6.6 becomes

$$\alpha = \left(2^k\right)^{\frac{1}{4}}$$

where 2^k is the factorial portion of design (number of factorial runs).

Using two factors, A and B (k = 2), CCD is illustrated as shown in Table 6.10.

Table 6.10 uses the concepts from Figure 6.17 and Equation 6.6 to develop the two full factorial central composite designs. Note that runs from 1 through 4 are

Table 6.10

A Two-Factor Composite Design (Coded Variables) of Cooling Time and Pack Pressure

Trial run	A (x_1)	B (x_2)
1	−1	−1
2	+1	−1
3	−1	+1
4	+1	+1
5	$-\alpha = 1.414$	0
6	$+\alpha = 1.414$	0
7	0	$-\alpha = 1.414$
8	0	$+\alpha = 1.414$
9	0	0
10	0	0

corner points, 5 and 6 are center replicates, 7 through 10 are star points $(0, \pm\alpha)$, $(\pm\alpha, 0)$, and 11 through 12 are star points for interaction.

Actual Values (uncoded values)

$$= \left(\frac{High + Low}{2} \right) + \left(\frac{High - Low}{2} \right) \times Coded \qquad (6.8)$$

The completed data set is shown in Table 6.11.

Using Equation 6.7 and Table 6.10, one can calculate the actual values of CCD prior to carrying out the experiment. Then, actual values from Table 6.11 will be used in the experiment to obtain the response (part diameter). The response from the experiment will be analyzed manually or using statistical software (in this case Minitab was used) in achieving an estimated regression coefficient for prediction model (y), analysis of variance (ANOVA) estimates. Thus, the results in Tables 4.12 and 4.13 were obtained using Minitab software.

Again, using the data from Tables 6.10, 6.11, and 6.12, information obtained by Minitab, the following model (prediction equation) is concluded:

$$y = 1.6650 - 0.00032x_1 + 0.00372x_2 + 0.00001x_1^2 + 0.00068x_2^2 - 0.0001x_1x_2$$

The stationary point is determined by differentiating y with respect to both variables and solving simultaneously for x_1 and x_2.

(a) $\dfrac{\partial y}{\partial x_1} = -0.00032 + 0.00002x_1 - 0.0001x_2$

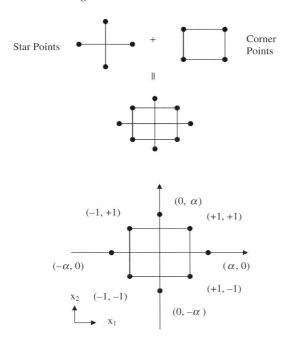

Figure 6.17   Central composite design for k = 2 (for k = 3, see Figure 10.2).

Table 6.11

A Two-Factor Composite Design (Uncoded Variables) of Cooling Time and Pack Pressure and Resultant Response Data

Trial run	A Cooling time (x_1)	B Pack pressure (x_2)	y Response (diameter)
1	12	10,000	1.6614
2	16	10,000	1.6623
3	12	14,000	1.6685
4	16	14,000	1.6690
5	11.172	12,000	1.6667
6	16.828	12,000	1.6639
7	14	9,172	1.6610
8	14	14,828	1.6723
9	14	14,000	1.6646
10	14	14,000	1.6653

Table 6.12

Minitab Design Criteria for CCD two-Level Full Factorial

Design criteria			
Factors:	2	Replicates:	1
Base runs:	10	Total runs:	10
Base blocks:	1	Total blocks:	1

Two-level factorial: full factorial

Cube points:	4		
Center points in cube:	2		
Axial points:	4		
Center points in axial:	0		

Alpha: 1.41421

Design table			
run	Blk	A	B
1	1	−1.00000	−1.00000
2	1	1.00000	−1.00000
3	1	−1.00000	1.00000
4	1	1.00000	1.00000
5	1	−1.41421	0.00000
6	1	1.41421	0.00000
7	1	0.00000	−1.41421
8	1	0.00000	1.41421
9	1	0.00000	0.00000
10	1	0.00000	0.00000

(b) $\dfrac{\partial y}{\partial x_2} = -0.00372 + 0.00001x_1 + 0.00136x_2$

Equating Equations (a) and (b) to zero results in

(a) $+0.00002x_1 - 0.0001x_2 = +0.00032$

(b) $-0.0001x_1 + 0.00136x_2 = -0.00372$

Multiplying Equation (a) by 0.0001 (coefficient of x_1 in Equation (b)) and multiplying Equation (b) by 0.00002 (coefficient of x_2 in Equation (a)), combining the results of both equations' outcome will come to

(a) $+2.0\text{E-}9x_1 - 1.00\text{E8}x_2 = +3.20\text{E8}$

(b) $\underline{-2.0\text{E-}9x_1 + 2.72\text{E8}x_2 = -7.44\text{E8}}$

    $0 \qquad\quad + 1.72\text{E8}x_2 = -4.24\text{E8}$

Table 6.13

Minitab Results for Central Composite Design two-Level Full Factorial Response Surface\Regression: y versus A, B

The analysis was done using coded units.
There are estimated regression coefficients for y.

Term	Coef	SE Coef	T	P
constant	1.66455	0.001825	912.038	0.000
A	0.00174	0.000913	1.910	0.129
B	0.00019	0.000913	0.211	0.843
A*A	0.00031	0.001207	0.259	0.809
B*B	−0.00019	0.001207	−0.155	0.884
A*B	−0.00010	0.001291	−0.077	0.942

S = 0.002581   R − Sq = 49.1%   R − Sq(adj) = 0.0%
Analysis of Variance** for y

Source	DF	Seq SS	Adj SS	Adj MS	F	P
Regression	5	0.000026	0.000026	0.000005	0.77	0.617
Linear	2	0.000025	0.000025	0.000012	1.85	0.270
Square	2	0.000001	0.000001	0.000001	0.08	0.927
Interaction	1	0.000000	0.000000	0.000000	0.01	0.942
Residual Error	4	0.000027	0.000027	0.000007		
Lack-of-Fit	3	0.000027	0.000027	0.000009	1776.17	0.017
Pure Error	1	0.000000	0.000000	0.000000		
Total	9	0.000052				

Obs	Std Order	y	Fit	SE Fit	Residual	St Resid
1	1	1.661	1.663	0.002	−0.001	−0.78
2	2	1.669	1.666	0.002	0.002	1.38
3	3	1.662	1.663	0.002	−0.001	−0.59
4	4	1.669	1.667	0.002	0.002	1.58
5	5	1.665	1.663	0.002	0.002	1.13
6	6	1.665	1.668	0.002	−0.003	−1.92
7	7	1.664	1.664	0.002	−0.000	−0.25
8	8	1.664	1.664	0.002	−0.001	−0.54
9	9	1.665	1.665	0.002	−0.000	−0.03
10	10	1.665	1.665	0.002	0.000	0.03

Estimated regression coefficients for y using data in uncoded units

Term	Coef
Constant	1.66455
A	0.00174268
B	0.000192678
A*A	0.000312500
B*B	−1.87500E-04
A*B	−1.00000E-04

**See Chapter 8 for details.

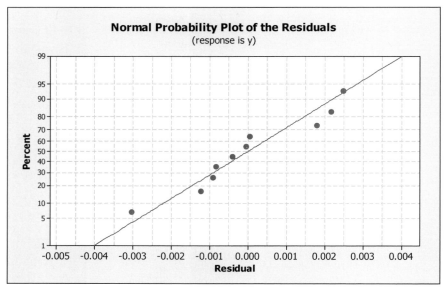

Figure 6.18   The response surface for the part diameter as a function of plastics cooling time (A) and plastics pack pressure (B), where the predicted diameter $y = 1.6650 - 0.00032\ x_1 + 0.00372\ x_2 + 0.00001\ x_1^2 + 0.00068\ x_2^2 - 0.0001\ x_1 x_2$.

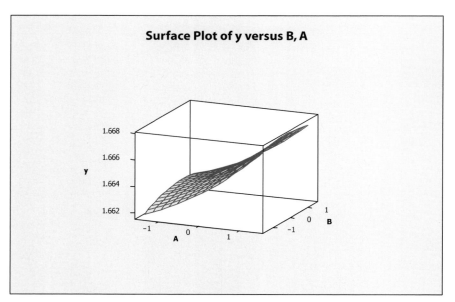

Figure 6.19   Normal probability of residual for response y.

Solve for x_2 $x_2 = -2.465$
Substitute x_2 in Equation (a) or (b) to solve for x_1
Taking Equation (b)

$$-0.0001x_1 + 0.00136(-2.465) = -0.0037$$
$$1.0E\text{-}4x_1 = 3.676E\text{-}4$$
$$x_1 = 3.676$$

where $1.0E - 4 = 1.0 \times 10^{-4}$

To test the answer for correctness, substitute x_1 and x_2 values in Equation (a) or (b). Equation (b) is tested here:

$$+0.00372 + 0.00136(-2.465) - 0.0001(3.676) = 0$$
$$0 = 0$$

Therefore, the stationary point is estimated to be at $(x_1, x_2) = (3.676, -2.465)$. The response surface and normal probability of residuals are given in Figures 6.18 and 6.19.

Chapter 7

Roles and Responsibilities to Lean *Six Sigma* Philosophy and Strategy

7.1 THE ROAD MAP TO LEAN *SIX SIGMA* PHILOSOPHY AND STRATEGY

Once company executive members accept the concept, companywide training will begin and may take from two to three years. Furthermore, companywide Lean *Six Sigma* achievements may take five or more years. This is for conversion from 3 sigma to 4, 4.5, 5, 5.5, and *Six Sigma*. However, each year that the sigma level increases, the company productivity improves, and profitability improves as well. Moreover, the challenge of attaining a higher sigma level also becomes tougher. All can be achieved with the right direction, determination, confidence, the right tools, and a timeline.

7.2 CREATION OF *SIX SIGMA* INFRASTRUCTURE

Six Sigma methodologies require full integration of all departments in the entire company; otherwise, the philosophy will not work. Managers cannot waive their commitment. The training should create four different levels of workforce: Champion, Master Black Belt (MBB), Black Belt (BB), and Green Belt (GB) or project leader. These titles originated from Motorola, but some companies have initiated their own titles.

Essentials of Lean *Six Sigma*

7.2.1 EXECUTIVE SPONSOR

As emphasized in Chapter 6, the executive sponsor should communicate, lead, and direct the company's overall objectives toward successful and profitable Lean *Six Sigma* implementation. The executive leadership has to inform the team that he or she is the driving force/sponsor and committed to implementing the Lean *Six Sigma* companywide. He or she will support the team and the program in every step. The *Six Sigma* program will not be successful without the direct involvement of executive leadership. CEO and executives must believe in the power of Lean *Six Sigma*. If they don't, then the *Six Sigma* program will fail for sure.

7.2.2 CHAMPION

The Champion should be from the executive members of the company. The members of this team could also include executive personnel or any other individual from upper-level management. This individual should possess managerial and technical skills in reinforcing, planning, allocating resources, and providing the necessary tools. In addition, he or she will be responsible for overseeing the results and success of the projects, selecting individuals to be trained as Master Black Belt and Black Belts with leadership to implement the world class concept and philosophy and more.

In larger corporations, Champions are divided among the business units. The corporate Champion reports to the CEO or president, and other business unit Champions report to the corporate Champion. In some corporations the CEO or president acts as a corporate Champion. Different corporations give Champions different names, such as quality leader, *Six Sigma* leader, or *Six Sigma* Champion. Further, responsibilities of the unit Champion are that he or she will ensure that the assigned financial objectives are achieved. He or she will obtain the reports from the controller to evaluate the impact of projects with outcome results before reporting to the corporate Champion as shown in Figure 7.1.

7.2.3 MASTER BLACK BELT

The Master Black Belt (MBB), also known as the quality manager, possesses knowledge of advanced applied statistical analysis (analysis of variance, design of experiment, response surface), business strategies, leadership training, and an extensive background in applying Lean *Six Sigma* methods. The MBBs are highly skilled in Lean *Six Sigma* techniques, and they mentor and teach the Black Belts. They must complete intensive training and oversee many projects before they

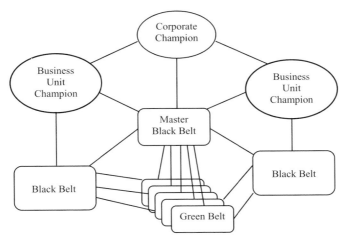

Figure 7.1 Organization of multiple Champions, Black Belts, and Green Belts.

earn certification. This is a hands-on, full-time position. Normally, the Master Black Belt is selected from among the top-talented Black Belts in the company. The MBB also performs as a consultant to BBs in their projects to help, push, or direct if any hangups need to be resolved or cleared. Nevertheless, the company expects the MBB to be expert in advanced tools and management of Lean as well as *Six Sigma* with already proven projects. The MBB should have completed several BB projects within a year.

7.2.4 BLACK BELT (TEAM LEADER)

This individual is the technical leader of the Lean *Six Sigma* project. The Black Belt should take a one-month training course and must be assigned to the Lean *Six Sigma* project for at least a couple of years. The course content has to be carried out with hands-on projects, each lasting approximately three to six months, with year 2005 standard savings in $350,000 to $500,000. The Black Belts, just like the Master Black Belts, should have demonstrated leadership and communication skills in their work assignments before being assigned to the training program. In addition, these individuals should possess the following capabilities:

- Mentoring other leaders in achieving the Lean *Six Sigma* goals.
  - Selling the idea and philosophy of the concept.
  - Developing, setting a direction, and leading the team to a higher level.

- Teaching, training, and coaching the Lean *Six Sigma* tools, as well as new techniques, case studies to project leaders in groups, and one-on-one cases.
  - Must have the ability to convert the concept to a highly successful project. They should pass the Lean *Six Sigma* skills and tools to their peers and to the customer-oriented team.
  - Developing an in-depth knowledge of Lean *Six Sigma* statistical tools (see Table 8.4) and techniques to improve key processes.
  - Developing and create techniques and shortcuts to achieve objectives.
- Keeping in contact with other organizations about the *Six Sigma* tools.
- Selecting project individuals who have high discipline in their area of work and are hands-on active with the process of production.
- Should be able to shift from one project to another and support the project leaders throughout the company.
- Be able to think like management (e.g., time, money, performance, and organizational dynamics).

7.2.5 GREEN BELT (TEAM PARTICIPANT)

The Green Belts work with Black Belts to solve problems. The individuals must have statistical knowledge and be trained in basic *Six Sigma* concepts, such as problem solving, statistical analysis, and so forth. Thus, they should be involved with Lean *Six Sigma* projects before obtaining the certification. Normally, the training period for a Green Belt is one week, which provides overview concepts and strategies in problem solving. Statistical tools are available in Minitab software (one of the many) for training.

7.2.6 TEAM RECOGNITION/COMPENSATION

Recognition of BBs and team members is important as motivation and a driving force for a team in achieving the objective of the project. Different companies have different programs in recognizing their people assets (i.e., promotion, stock share, some percent of base salary profit share, etc.).

Chapter 8

Road Map to Lean *Six Sigma* Continuous Improvement Engineering Strategies

8.1 *SIX SIGMA* CONTINUOUS IMPROVEMENT ENGINEERING

In reference to our earlier discussion of process quality in Chapter 1, it is evident that as the sigma level approaches six, the defects of a process drop to 3.4 defects per million parts. The success of Japanese manufacturers comes from focusing on *process quality* rather than *part quality*. In most cases a defect-free process produces a defect-free product. Therefore, a quality product requires a quality process. Nevertheless, all the defects are the result of improper design, process condition, material, or machine operation. Design for Lean *Six Sigma* (DFLSS) will evaluate all the above factors in achieving organizational goals. Companies that are willing to implement Lean *Six Sigma* should recognize the following criteria:

- Top management should engage in the program.
- The company needs to be committed to providing resources for continuous improvement.
- The company needs to believe in Lean *Six Sigma* capabilities in achieving and solving very difficult problems.
- The company needs to be committed to customer satisfaction and bottom line.

The following sections examine how to engineer *Six Sigma* continuous improvement, Design for *Six Sigma* (DFSS), and Lean *Six Sigma*. *Six Sigma* process improvements consist of five different phases: Phase 0—Process Definition/ Project Selection; Phase I—Process Measurement; Phase II—Process Analysis;

Phase III—Process Improvement; and Phase IV—Process Control and Maintain.

8.2 DEFINITION AND MEASUREMENT

8.2.1 PHASE 0: PROCESS DEFINITION/PROJECT SELECTION

Project Objectives and Strategies

This step focuses on the following objectives and actions:

1. Identify business financial drivers—the area that will bring maximum dollars by preventing or eliminating waste.
2. Determine critical-to-quality processes through voice-of-customer (VOC) requirements. Define customer specification limits (e.g., LSL and USL) and any other requirements.
3. Define project issues, specify project goals and values, such as:
   - Select a project (if possible with maximum number of variations) with criteria of resources needed, expertise available, process complexity, likelihood of success, and support or buy-in.
   - Identify process parameter input (independent or causes) and output (dependent or symptoms) variables.
   - Establish the process flowchart.
   - Collect data on process response.
4. Set boundaries of process or voice-of-process (VOP)—for instance, what is included and what is excluded.
5. Select fully trained project team members. The success of the *Six Sigma* project will be obtained by giving responsibilities to those who are actually involved in the process.
6. Establish execution plan.

Six Sigma follows the projects in which solutions are not known. It is about solving a business problem by improving processes. A *Six Sigma* project should include at least one of the following criteria:

1. Process with lower sigma level.
2. Return on investment (ROI). The higher the ROI, the higher the impact.
3. Quantitatively measurable.
4. High probability of success.
5. Be able to complete within three to six months
6. High on the corporate priority list or an important project to the organization.

The results on business should have:

1. Impact on business strategy and competitive position.
2. Impact on external customers and requirements.
3. Impact on "core competencies."
4. Financial impact.
5. Urgency.
6. Trend.
7. Sequence or dependency.

Hints: Where to look for such a project?

- *Internal defects*

  These opportunities are available in the area of scrap (defect products) reduction, parts, materials, rework (keep accurate track of rework by defects)/recycle in-house, reinspection, staff overtime, costs, excess labor or labor-intensive projects, staff turnover, absenteeism, and anything one has a Poka-Yoke for. One should know how often the Poka-Yoke is catching something and then try to prevent it rather than catch it.

  Furthermore, the internal defects include higher cycle time, higher lead time, and wait time, excess inventory, transportation, machine and automation downtime, cost reduction with financial impact, nonvalue-added activities, and large projects with larger budgets (benchmark certain applications for cost reduction). Avoid projects with no value-added or financial impacts on the bottom line.

- *External defects*

  Opportunities are available in areas such as warranty returns, product recalls, field failures, and anything that has been audited or rejected for either being out-of-specification or not meeting regulations.

Identifying Process Variation

The following are question in understanding the variation. How much variation is there in incoming materials and how does this affect the output? Can one scientifically adjust the process to compensate for changing material or air temperature variation? Where is input required to be controlled to continuously get an acceptable output always?

General Guidelines for Project Selection

- A project should have identifiable process inputs and outputs and follow Equation 3.3: (Output = f (input)).

- For projects that have operator or operator training as an input, focus on ways to reduce operator variation—hence making the process more robust to different or untrained operators.
- A *Six Sigma* project should never have a predetermined solution. If one already knows the solution, then the problem should be fixed!
- All projects are required to be approached from the aspect of identifying the variation inputs, controlling them, and removing the defects.

EXAMPLE 8.1.   **Issue:** A company X is experiencing longer cycle time at station 45 due to receiving bad parts from station 35, and they must rework them.

Old method solution: Rebalance the line in order to do the rework and keep cycle time below customer specifications while not spending extra labor cost.

New method (Six Sigma) solution: Investigate and control key inputs that contribute to producing a bad part at station 35.

EXAMPLE 8.2.   **Issue:** A company Y has had two quality issues reported this year for missing armrest screws.

Old method solution: Add sensors to detect screws further down the line. If screws are missing, operator manually fixes the problem.

New method (Six Sigma) solution: Find process inputs (Xs) causing missing screws. For instance, auto gun does not always feed correctly due to air pressure variation. Either study ranges required for 100% operation and control in that range, or find a way to make the auto gun more robust to the range of variation experienced.

Process Definition Tools and Techniques

The management and planning tools are described in Phase 0 (define) through Phase IV (control and maintain) as follows:

- The affinity diagram is used in the following:
  - Refine breakthrough thinking (to generate a high volume of ideas or issues).
  - Tap the creative side of the brain (to capture ideas not available in a typical meeting).
    - Creatively brainstorm and organize a large number of ideas.
    - Summarize the natural groupings among ideas to understand the essence of a problem.
  - Brainstorm ideas for the issues, solutions, or problems.
    - State the question, record ideas, review ideas, prioritize.

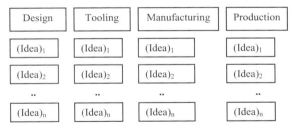

Figure 8.1   Affinity diagram.

Table 8.1

Examples of an Affinity Diagram

Part design	Tool design	Plastics molding	Product productivity
Material selection	Steel selection	Cycle time	Equipment
Material shrinkage	Cooling system	Pressure	Reliability
Material property	Sharp corner	Speed	Training
Wall thickness	Gate size	Temperature	Shift
Functionality	Runner size	Shot size	Absenteeism
Tolerance	Texture	Cooling time	
Draft	Hot runner	Machine size	

- Sort and classify ideas in at least five through ten different groups.
- Create a summary for each group.

Figure 8.1 illustrates these concepts.

Table 8.1 illustrates industrial examples of the Figure 8.1 affinity diagram.

- Interrelationship Diagraph
  - Show graphical representation of all the factors in a complicated problem, system, or situation.
  - Identify, analyze, and classify the cause-and-effect relationship matrix between issues.
  - Determine key drivers or outcomes for effective problem resolutions.
  - Draw final interrelationship diagram and identify the largest factors (drivers) and outcome.

Factors effecting production for injection molding parts are shown in Figure 8.2.

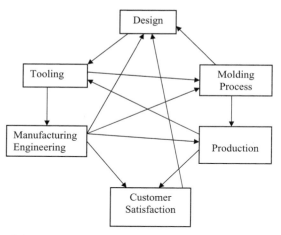

Figure 8.2   Interrelation diagraph for injection molding manufacturing.

- Quality Function Deployment (QFD)
  House of quality or QFD is a communication instrument that provides
  focus to the best opportunities. It is a priority of actions based on a
  matrix that uses requirement limits in the design to achieve an excellent
  product. It consists of the following matrices in five steps:
  1. Relationship matrix of *how(s)* (company measurements) and *what(s)*
     (customer requirements and priority), as shown in Figure 8.3 and
     described in the following two categories:
     a. Technical *whats*
        i. What does the customer need and want?
        ii. What are the customer's priorities and importance ratings?
     b. Technical *hows*
        i. Tools to achieve the goals of *whats*
        ii. Direction for improvement and how to get to goals
        iii. Tool: tree diagram structure used to document the subgroup of
             hows
        iv. Relationship matrix depends on
            - How will each *how* fulfill each *what*?
            - Two-stage analysis
            - Is there a relationship?
            - If yes, is it a low (score = 1), medium (score = 3), or high
              (score = 9)?
  2. Correlation matrix
     a. Illustrates positive and negative relationships among the *hows*
     b. Focuses on complementary and conflicting *hows*

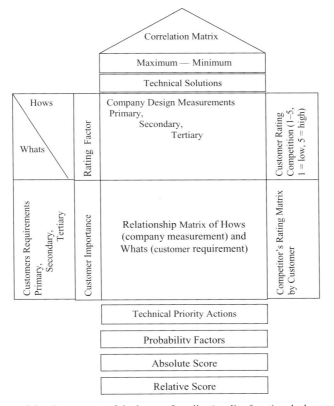

Figure 8.3   Components of the house of quality (quality function deployment).

3. Competitor's rating matrix by the customers
   a. Investigates competitive benchmarking on *whats*
   b. Evaluates weaknesses and strength of products
   c. Finds out what the competitors' products lack
4. Technical assessment
   a. Investigates competitive benchmarking on *hows*
   b. Evaluates weaknesses and strengths of product or service
   c. Finds out about the drawbacks in competitors' product or service
   d. Achieves targets and specification to outperform the competitors' market
   e. Applies safety and ergonomic requirements
5. Prioritizing customer need (important rating factors)
   a. Accurately reflect customer opinions
   b. Weighting factors

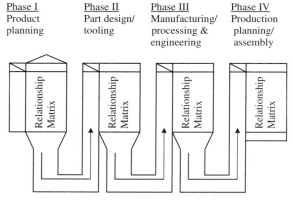

Phase I	Phase II	Phase III	Phase IV
Product planning	Part design/ tooling	Manufacturing/ processing & engineering	Production planning/ assembly

Figure 8.4   Four phases of QFD.

c. Simple scale
  i. 1 to 10 (1 = minor issue and 10 = serious issues)
  ii. 1 to 5 (1 = minor and 5 = serious)
d. Absolute or relative score (see Chapter 9 calculation formula)

Full QFD for all steps of product development in four phases (product planning, part design/tooling, manufacturing/processing and engineering, production planning/assembly) is illustrated in Figure 8.4.

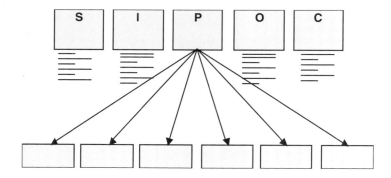

- Overall Business Process Map (SIPOC)
  SIPOC (or **S**upplier, **I**nputs, **P**rocesses, **O**utputs, and **C**ustomer) objectives are as follows:
  a. To identify all relevant elements of a process improvement project before it begins, such as suppliers, inputs, process, outputs, as well as customers of the processes (SIPOC), and their requirements. First, to establish SIPOC, one may list and sequence the steps.

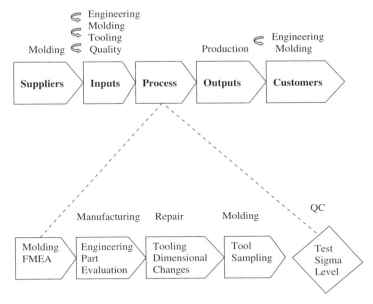

Figure 8.5   SIPOC for tool repair.

b. To distinguish process boundaries (X = process inputs and Y = process outputs).
c. To map data collection sources. Validate and finalize the flowchart.
d. To identify opportunities for improvement.

The SIPOC example shown in Figure 8.5 is a mold repair process in the plastics injection molding processing industry.

- Process Mapping
  Process mapping is a process flowchart, which displays an accurate and detailed picture of the process. It also helps to explore more improvements—that is, SIPOC, FMEA, C_p, C_{pk}, P_p, P_{pk}, and so on.
- Project Charter
  Project charter concentrates on the following:
  a. *Six Sigma* scorecard: Steps completed should be signed by project sponsor (see Appendix for an example)
  b. Project title
  c. Project timing and deadlines
  d. Project scope
     i. Problem statement
     ii. Desired goal statement
  e. Milestones

8.2.2 PHASE I: PROCESS MEASUREMENT

Thorough measurement of the existing process is necessary for redesign or future designs of process/product. The existing process should be examined without any manipulation. Data collected (continuous or discrete) will represent the actual real-time capability of the process for the products under production. For instance, data collected in a production process should be relevant to output variables (dependent or response) such as dimensions, mechanical properties (impact, tensile strength, and quality), as well as input variables (manipulated) of the machine—for example, pressure, volume, temperature, velocity, screw recovery time, plastics concentration ratio, plastic shot size, and so on. The data should be selected randomly with a large sample size or population.

Process Measurement Objectives

Highlights of the process measurement procedure are the following:

a. Develop and define key process measures and clarify goals.
b. Identify the vital few processes that have the greatest impact on the project.
   i. Identify process parameters [process input X's (independent) and process output Y's (dependent)] and correlations between them:

$$((Y = f(x)).$$

   ii. Create a detailed process map (SIPOC of existing process) to identify possible measures. See Section 8.2.1.
c. Collect and analyze data.
   i. Collect data on key process response and convert any transactional data to numerical data.
   ii. Develop operational definitions and procedures.
   iii. Validate the measurement system.
   iv. Start data collection and continue to improve the measurement consistency.
d. Measure performance (output) variables.
e. Conduct a process capability study (C_p, C_{pk}, P_p, P_{pk}).
f. Review probability and statistical tools.
g. Implement the five *whys* method (see Section 8.4.3).
h. Continue to ask five *whys* that eventually will lead to the root cause of the problem.

Measurement Tools and Techniques

Some of the important tools and methods to achieve measurement objectives are as follows:

- Develop process maps using the flowchart and deployment chart of the existing process.
- Clarify process inputs (X's) and process output identification.
- Establish a data collection plan, design check sheet as shown below:

	Day 1	Day 2	Day 3	Day 4
Paint	1111	111	1	11111
Test	1	11111	111	111
Perf	111	111	1111	111

Collect data about how or what type of problem is occurring and distinguish between opinion and fact. For instance, agree on the definition of the observations and determine who will collect, what will be collected, and when it will be collected. Then, design the check data sheet and start the process of collecting.

- Pareto charts
Construct and prioritize the defect factors by adopting Pareto charts and their analysis. A Pareto chart is a column chart and is used to prioritize the problem-solving order so it will determine which issue has the most effect in the process. The Pareto chart applies to all the frequencies (outputs or symptoms and/or defects) of any process in any organization. Furthermore, it also does the following:

- Identifies the 20% of sources that causes 80% of the problems.
- Concentrates on the "vital few (most dominant) factors."
- Basically, the Pareto chart is another form of histogram with the only difference being that within the Pareto chart the frequencies start from highest to lowest.
An example of a Pareto chart in tool repair is shown here (the cost of tool repair is a function of mold numbers). The Excel spreadsheet or any other statistical software can be used to construct a Pareto chart.

Damaged Core/Cavity Cost Function of Mold

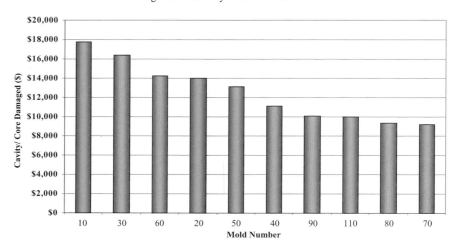

- Histogram
  Histogram illustrates frequency data in the form of a bar graph with
  bell-shaped looks. It is a highly effective tool in identifying the mean and
  capability of the process. Normally the frequency represents the
  dependent variable (y-axis) and independent variable (x-axis).

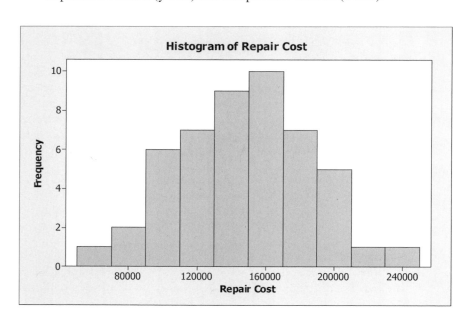

a. Construct and interpret a histogram—bell shaped, binomial, skewed, truncated, isolated peak, comb, and so on.
b. Summarize data variation in the process.
c. Visually and graphically communicate process data.
d. Apply process capability tools (C_p, C_{pk}, P_p, P_{pk}) and calculations.

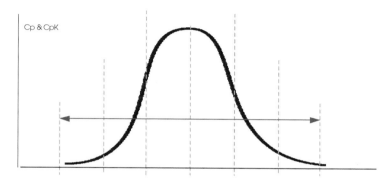

- Control Charts
  Construct and interpret control charts for the existing process using:
  a. Continuous data: obtained by the use of a measuring system and plotted in the following charts (also see the "Process Control Tools and Techniques" section for more details)
      i. X-bar ($\overline{X}$) and R charts: applied to continuous processes—that is, time, weight, and height. In the plastics industry the most common process is the extrusion process (continuous with respect to time). The X-bar chart monitors time difference, and S monitors variation.
      ii. $\overline{X}$ and S charts: similar to $\overline{X}$ and R charts with the only difference being that sigma is used as a dispersion of measurement.
      iii. X and mR charts: applied for long intervals observations.
  b. Discrete data: This is referred to as attributes, counts, percentages, and ordinal data. Percentages are equal to the proportion of items with a given characteristic and the need to count both occurrences, and nonoccurrences. Occurrences must be independent. For count data, it is impossible or impractical to count a nonoccurrence, and the event must be rare. The following is included in the charts:
      i. nP charts: applied to track the proportions of defective units when the sample size is constant.
      ii. P charts: similar to nP chart with sample size varying.
      iii. C charts: applied to samples with more than one defect per unit.
      iv. U charts: applied to track defects per varying size.

- Gage R&R: A study of variability in a product or process.
  - Repeatability: someone taking the same measurement (more than once) on the same item with the same instrument (gage) will get the same answer (normally more than ten parts maximum).
  - Reproducibility: different people measuring the same item with the same instrument (gage) will get the same answer.
- Sample sizing/data type.
  Sampling approaches are random, stratified random, systematic, and subgroup sampling. They are defined as follows:
  - *Random*—each unit has the same chance of being selected.
  - *Stratified random*—randomly sample a proportionate number from each group.
  - *Systematic*—sample every nth one (e.g., every fourth or tenth, etc.).
  - *Subgroup*—sample n units every fourth time (e.g., 3 units every hour) and then calculate the mean for each subgroup.
  - Continuous data: based on continuum of numbers—for example, currency, time, weight, and height. Sample size can be calculated using Equation 8.1.

$$n = \left(\frac{zS}{E}\right)^2 \tag{8.1}$$

where n is the sample number, z the standard normal distribution (z score), sample standard deviation, and E the measurement error.
  - Discrete (attribute/discontinue) data: based on counts or classes—for example, flipping a coin to yield a head or a tail. Sample size can be calculated using Equation 8.2.

$$n = \frac{(z^2)pq}{E^2} \tag{8.2}$$

where $q = p - 1$, p is the estimated proportion of correct observation, and q is the estimated proportion of defects (scraps). Then Equation 8.2 becomes 8.3.

$$n = \frac{(z^2)p(p-1)}{E^2} \tag{8.3}$$

Since *Six Sigma* is a statistical measurement, it follows that accuracy of process measurement is highly important in *Six Sigma*. In this step the nature of the existing process will be identified. The outcome will represent the actual real-time capability of the process for the products under current production. Highlights of the objectives and tools are in the following sections.

Data Stratification

Data stratification divides data into groups. It also detects a pattern that localizes a problem or explains why the frequency of impact varies between time, location, or conditions. The ways to stratify are who, what, where, and when.

Key Basic Statistical Backgrounds

One can identify the process mean (μ), sigma range (σ), and defect percent per million, as well as machine and process capability (C_p, C_{pk}, P_p, P_{pk}) for the existing manufacturing performance to create flowcharts, Pareto charts, and histograms. This can be done through statistical strategies given in the foregoing discussion or any other statistical resources. For further detailed review, refer to any statistical texts. On the other hand, perhaps it would be faster if one uses a statistical software tool such as Microsoft Excel with Macros, which has basic statistical options, and/or Minitab. Furthermore, any spreadsheet software with proper formulations will do the same calculation. It is necessary to learn basic statistical tools as listed in Table 8.4. Due to the importance of the capability study in *Six Sigma*, the succeeding discussion will review the mean, sigma, and process capability that are required for the measurement phase.

Arithmetic Mean

A population mean is the sum of the entire set of objects, observations, or scores that has something in common. This can be expressed as Equation 8.4.

$$\mu = \sum_{i=1}^{N} X_i / N \qquad (8.4)$$

where
μ = population mean
X_i = Part measurement
N = Total number of parts (population)

A sample is a set of observations drawn from a population. Since it is usually time consuming and too costly to test every one of a population, a sample from the population is normally the best approach to ascertain the population outcome. Sample mean is expressed in Equation 8.5.

$$\bar{x} = \sum_{i=1}^{n} x_i / n \qquad (8.5)$$

where
$\bar{x}$ = Sample mean
x_i = Sample measurement
n = Total number of sample parts

Population Variance and Standard Deviation

The variance Equation 8.6 is a measure of variability (spread out) in a population data set or distribution curve.

$$\sigma^2 = \frac{\sum_{i=1}^{N} (X_i - \mu)^2}{N} \tag{8.6}$$

and mathematically the standard deviation Equation 8.7 is the square root of variances

$$\sigma = \sqrt{\frac{\sum_{i=1}^{N} (X_i - \mu)^2}{N}} \tag{8.7}$$

where, σ^2 is the variance and σ is the standard deviation of the population.

Sample Variance and Standard Deviation

The variance Equation 8.8 is a measure of variability (spread out) in a sample data set or distribution curve.

$$S^2 = \frac{\sum_{i=1}^{n} (x_i - \bar{x})^2}{n} = \frac{\sum x^2 - \frac{(\sum x)^2}{n}}{n-1} \tag{8.8}$$

And sample standard deviation Equation 8.9 is the square root of variances

$$S = \sqrt{\frac{\sum_{i=1}^{n} (x_i - \bar{x})^2}{(n-1)}} \tag{8.9}$$

where S^2 is the variance and S is the standard deviation of the sample. (The accuracy of standard deviation and normal distribution increases with the higher sample size.)

Hypothesis Testing

Hypothesis testing is a method used in inferential statistical analysis and related to confidence intervals. By definition, if an inspector claims that the mean of a population is equal to $\mu = a$, then, the fact is that either $H_0 : \mu = a$ or $H_a : \mu \neq a$. So one needs to test and determine to reject $H_0 : \mu = a$ or fail to reject H_0, where H_0 is called null hypothesis, which the investigator wishes to discredit that

$(H_0 : \mu = a)$, and H_a is called alternative hypothesis, which the investigator wishes to support that $(H_a : \mu \neq a)$. The following phases are the techniques for hypothesis testing.

1. State the problem conditions and focus on the test—for instance,
   a. Null hypothesis $H_0 : \mu = \mu_o$
   b. Alternative hypothesis $H_a : \mu \neq \mu_o$
   Note: The conditions for samples with z-test and population mean are as follows:
   a. The normal population or large samples $n \geq 30$ and σ is known.
   b. If the tests are two tails, then alternative H_a has a sign of $(H_a : \mu \neq \mu_o)$.
   c. If it is the left tail, H_a has a sign of less than $(H_a : \mu < \mu_o)$.
   d. If it is the right tail, then H_a has a sign of greater than $(H_a : \mu > \mu_o)$.
2. Define the critical values and test statistic—that is, for testing large samples $n \geq 30$, use z-Table (or t) and define the rejection region.

$$Z = \frac{(\overline{X} - \mu_o)\sqrt{n}}{\sigma} \approx \frac{(\overline{X} - \mu_o)\sqrt{n}}{s} \tag{8.10}$$

3. Calculate the value of the test statistic (Equation 8.10).
4. Determine whether to reject H_0 or fail to reject H_0.
5. Report a conclusion in terms of the original problem.

The preceding discussion is detailed in Example 8.3.

EXAMPLE 8.3.   Recall the lightbulb reliability concept from Example 3.3 with number of samples 100 and average lifecycle of 400 hours, with the standard deviation of 39 hours. What conclusion can you make using a significance of $\alpha = 0.10$?

Use the preceding steps:

1. Define the hypothesis test conditions:

$$H_0 : \mu = 385$$
$$H_a : \mu \neq 385$$

2. State z-test statistic.

$$Z = \frac{(\overline{X} - \mu_o)\sqrt{n}}{\sigma} \approx \frac{(\overline{X} - \mu_o)\sqrt{n}}{s}$$
$$= \frac{(\overline{X} - 385)\sqrt{100}}{\sigma} \approx \frac{(\overline{X} - 385)\sqrt{100}}{s}$$

3. Determine the rejection region (see Figure 8.6).

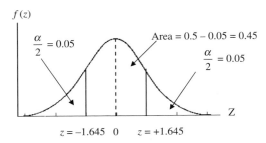

$f(z)$

$\dfrac{\alpha}{2} = 0.05$

Area $= 0.5 - 0.05 = 0.45$

$\dfrac{\alpha}{2} = 0.05$

$z = -1.645 \quad 0 \qquad z = +1.645$

Figure 8.6   Normal distribution curve for the Example 8.1.

Reject if z > 1.645 or z < −1.645

4. Determine the z-value.

$$Z = \frac{(400 - 385)\sqrt{100}}{39} = 3.85$$

Since z-value 3.85 > 1.645, then the conclusion is to reject H_0. In Figure 8.6, z-test falls in the rejection area.

5. Conclusion: According to sample data, there is enough proof to conclude that the average lifecycle of lightbulbs is not 385 hours.

Analysis of Variance (ANOVA)

Analysis of variance is a technique for testing the hypothesis about the mean. It is a frequently used science of statistical inference technique for analyzing the output of experimental data or transactional data in the form of mathematics.

One-Way ANOVA

The one-way (also called the one-factor) ANOVA normally states that a hypothesis test can be used in determining if the mean of one or more populations is different or not based on one independent factor. These are the requirement steps:

1. Independent samples from each population.
2. Samples from each population follow normal distribution with the same standard distribution (σ) and common variance (σ^2).
3. $H_0: \mu_1 = \mu_2 = \mu_3 = \ldots = \mu_{m-2} = \mu_{m-1} = \mu_m$
   H_a: not all means are equal. This is always a right tail test.
   Note: H_a defines that at least two of the means (μ's) are different or unequal.
4. Apply α value.
5. Establish one-factor ANOVA table (see Table 8.2).

Table 8.2

One-Factor ANOVA (Equations 8.11–8.14)

Source of inconsistency (variation)	Degrees of freedom (df)	Sum of squares (SS)	Mean squares (MS)	Statistical test (F)
Factor	$k - 1$	SS (*factor*)	$MS\,(factor) = \dfrac{SS\,(factor)}{k-1}$	$F = \dfrac{MS\,(factor)}{MS\,(error)}$
Error	$n - k$	SS (*error*)	$MS\,(error) = \dfrac{SS\,(error)}{n-k}$	
Total	$n - 1$	SS (*total*)		

$$MS = meansquare = (ss)(df)^{-1} \qquad (8.11)$$

Sum of squares:

$$SS\,(factor) = \left(\frac{T_1^2}{n_1} + \frac{T_2^2}{n_2} + \frac{T_3^2}{n_3} + \cdots + \frac{T_{m-2}^2}{n_{m-2}} + \frac{T_{m-1}^2}{n_{m-1}} + \frac{T_m^2}{n_m} \right) - \frac{T^2}{n}$$

$$n = n_1 + n_2 + n_3 + \cdots n_{m-2} + n_{m-1} + n_m$$

$$T = T_1 + T_2 + T_3 + \cdots + T_{m-2} + T_{m-1} + T_m = \sum x$$

(8.12)

$$SS(total) = \sum x^2 - \frac{T^2}{n} \qquad (8.13)$$

$$SS\,(error) = SS\,(total) - SS\,(factor)$$

$$= \sum x^2 - \frac{T^2}{n} - \left[\left(\frac{T_1^2}{n_1} + \frac{T_2^2}{n_2} + \frac{T_3^2}{n_3} + \cdots + \frac{T_{m-1}^2}{n_{m-2}} + \frac{T_{m-1}^2}{n_{m-1}} + \frac{T_m^2}{n_m} \right) - \frac{T^2}{n} \right]$$

$$= \sum x^2 - \left(\frac{T_1^2}{n_1} + \frac{T_2^2}{n_2} + \frac{T_3^2}{n_3} + \cdots + \frac{T_{m-2}^2}{n_{m-2}} + \frac{T_{m-1}^2}{n_{m-1}} + \frac{T_m^2}{n_m} \right)$$

(8.14)

Degrees of freedom:

$$df\,(factor) = k - 1$$

$$df\,(error) = n - k$$

$$df\,(total) = df\,(factor) + df\,(error) = k - 1 + (n - k) = k - 1 + n - k = n - 1$$

6. Make a conclusion:

Reject H_0 if $F_{calculated} > F_{\alpha, k-1, n-k}$

where, $F_{\alpha, k-1, n-k}$ is obtained from the F-Table in the Appendix.

m = Number of populations

n_k = Number of observations in the kth sample

k = Number of levels or treatment
n = Total number of observations

EXAMPLE 8.4.   In injection molding of plastics, the process temperature of plastic melt can be set at 280°F, 310°F, and 350°F. Determine if the temperature level affects the part weight significantly. Use significant level $\alpha = 0.05$.

The experiment ran with five trials at each level of temperature. Samples from 15 runs were collected randomly.

Temperature

280°F	310°F	350°F
12.50	12.00	11.00
13.00	11.75	10.25
11.70	11.00	10.00
14.10	11.25	10.00
12.75	11.50	11.50

The preceding table shows the part weights (y) information in grams. The three populations show that their sample means (μ) are not equal. A hypothesis test will indicate if there is a significant difference between them. The hypothesis test steps are as follows:

1. All three populations are obtained independently and randomly.
2. $H_0: \mu_1 = \mu_2 = \mu_3$
   H_a: Not all means are equal. This is always a right-tail test.
3. $\alpha = 0.05$
4. Temperature

	280°F	310°F	350°F
	12.50	12.00	11.00
	13.00	11.75	10.25
	11.70	11.00	10.00
	14.10	11.25	10.00
	12.75	11.50	11.50
Total (T):	$T_1 = 64.05$	$T_2 = 57.50$	$T_3 = 52.75$
	$y_1 = 12.81$	$y_2 = 11.45$	$y_3 = 10.55$

$n = 15$
Reading per treatment = 5
Number of levels = $m = 3$

Sum of squares:

$$SS(factor) = \left(\frac{T_1^2}{n_1} + \frac{T_2^2}{n_2} + \frac{T_3^2}{n_3} \right) - \frac{T^2}{n}$$

$$n = n_1 + n_2 + n_3 = 5 + 5 + 5 = 15$$

$$T = T_1 + T_2 + T_3 = \sum y = 174.30$$

$$T^2 = 30,380.49$$

$$\sum y^2 = (12.50)^2 + (13.00)^2 + (11.70)^2 + (14.10)^2 + (12.75)^2$$
$$+ (12.00)^2 + (11.75)^2 + (11.00)^2 + (11.25)^2 + (11.50)^2$$
$$+ (11.00)^2 + (10.25)^2 + (10.00)^2 + (10.00)^2 + (11.50)^2$$
$$= 2,043.70$$

$$SS(total) = \sum y^2 - \frac{T^2}{n} = 2,043.70 - \frac{30,380.49}{15} = 18.33$$

$$SS(factor) = \left(\frac{T_1^2}{n_1} + \frac{T_2^2}{n_2} + \frac{T_3^2}{n_3} \right) - \frac{T^2}{n}$$
$$= \left(\frac{64.05^2}{5} + \frac{57.50^2}{5} + \frac{52.75^2}{5} \right) - \frac{174.30^2}{15} = 12.87$$

$$SS(error) = SS(total) - SS(factor)$$
$$= \sum y^2 - \left(\frac{T_1^2}{n_1} + \frac{T_2^2}{n_2} + \frac{T_3^2}{n_3} + \cdots + \frac{T_{m-2}^2}{n_{m-2}} + \frac{T_{m-1}^2}{n_{m-1}} + \frac{T_m^2}{n_m} \right)$$
$$= \sum y^2 - \frac{T^2}{n} - \left[\left(\frac{T_1^2}{n_1} + \frac{T_2^2}{n_2} + \frac{T_3^2}{n_3} \right) - \frac{T^2}{n} \right]$$
$$= 2,043.70 - \left(\frac{64.05^2}{5} + \frac{57.50^2}{5} + \frac{52.75^2}{5} \right)$$
$$= 2,043.70 - 2,038.24$$
$$= 5.46$$

Degrees of freedom: Since $k = 3$ and $n = 15$, the (df)s are

$$df(factor) = k - 1 = 3 - 1 = 2$$
$$df(error) = n - k = 15 - 3 = 12$$
$$df(total) = df(factor) + df(error) = k - 1 + (n - k) = k - 1 + n - k = n - 1$$
$$= 15 - 1 = 14$$

$$MS(factor) = \frac{SS(factor)}{k - 1} = \frac{12.87}{2} = 6.44$$

$$MS(error) = \frac{SS(error)}{n - k} = \frac{5.46}{12} = 0.46$$

$$F = \frac{MS(factor)}{MS(error)} = \frac{6.44}{0.46} = 14$$

One-Factor ANOVA Table

Source of inconsistency (variation)	Degrees of freedom (df)	Sum of squares (SS)	Mean squares (MS)	Statistical test (F)
Factor	$k - 1 = 2$	SS (factor) = 12.87	MS (factor) = 6.44	F = 14.00
Error	$n - k = 12$	SS (error) = 5.46	MS (factor) = 0.46	
Total	$n - 1 = 14$	SS (total) = 18.33		

5. Make a conclusion.
   Reject H_0 if $F_{calculated} > F_{\alpha,k-1,n-k}$
   where, $F_{\alpha,k-1,n-k}$ is obtained from the F-Table in the Appendix.
   $F_{calculated} = 14$
   $F_{\alpha,k-1,n-k} = F_{0.05,2,12} = 6.93$ from the F-Table in the Appendix. This is a right tail test.
   Reject H_0 if $F_{calculated} > F_{\alpha,k-1,n-k} = 6.93$
   Because, $14 > 6.93$, null hypothesis ($H_0: \mu_1 = \mu_2 = \mu_3$) is rejected.

Therefore, the conclusion is that at significance level $\alpha = 0.05$, data show temperatures do have an impact on weight.

Two-Way ANOVA

As described, one-way ANOVA is used in analyzing experiments with one independent factor. The two-factor ANOVA can be applied in two-factor factorial design of experiments (DOE) or designs with more than one factor. In one-factor, analysis is based on only one single factor, with different levels with no interaction. The two-way method analyzes two factors with effects of factor one and two, as well as interactions. The goal is to find out if the individual factors (say A or B) have a significant or any effect on the output of the experiment. Consequently, there are three sets of hypotheses tests that can be applied to two-factor factorial design. The big difference between the one- and two-factor ANOVA procedure is in constructing the ANOVA as shown in Table 8.3.

Sum of the Squares

Factor A (background):

Table 8.3

Two-Factor ANOVA Table with Interaction (Consists of Five Sources of Variation)

Source of inconsistency (variation)	Degrees of freedom (df)	Sum of squares (SS)	Mean squares (MS)	Statistical test (F)
Factor A	$\alpha - 1$	SS (factor A)	$MS(factor.A) = \dfrac{SS(factor.A)}{\alpha - 1}$	$F_A = \dfrac{MS(factor.A)}{MS(error)}$
Factor B	$\beta - 1$	SS (factor B)	$MS(factor.B) = \dfrac{SS(factor.B)}{\beta - 1}$	$F_B = \dfrac{MS(factor.B)}{MS(error)}$
Interaction A × B	$(\alpha - 1)(\beta - 1)$	SS (Interaction AB)	$MS(Interaction.AB) = \dfrac{SS(Interaction.AB)}{(\alpha - 1)(\beta - 1)}$	$F_{AB} = \dfrac{MS(Interaction.AB)}{MS(error)}$
Error	$\alpha\beta(r - 1)$	SS(error)	$MS(error) = \dfrac{SS(error)}{\alpha\beta(r - 1)}$	
Total	$n - 1 = \alpha\beta r - 1$	SS(total)		

$$SS(factor.A) = \frac{1}{\beta r}\left(T_1^2 + T_2^2 + \cdots + T_m^2\right) - \frac{T^2}{\alpha\beta r} \tag{8.15}$$

where T = total of all $n = \alpha\beta r$ readings—that is, $T = T_1 + T_2 + \cdots + T_m$.

Factor B (guideline):

$$SS(factor.B) = \frac{1}{\alpha r}\left(S_1^2 + S_2^2 + \cdots + S_k^2\right) - \frac{T^2}{\alpha\beta r} \tag{8.16}$$

Interaction AB: $SS(Interaction.AB) = \dfrac{\sum R^2}{r} - \left[SS(factor.A) + SS(factor.B) + \dfrac{T^2}{\alpha\beta r}\right]$

$$\tag{8.17}$$

Where ΣR^2 sum of all the squares of replicates total:

$$\text{Total: } SS(\text{total replicate}) = \sum x^2 - \frac{T^2}{\alpha\beta r} \tag{8.18}$$

Where Σx^2 is the sum of the squares for each of the $n = \alpha\beta r$ observations. By subtraction,

$$SS(error) = SS(\text{total replicates}) - [SS(\text{factor A}) + SS(\text{factor B}) + SS(\text{Interaction AB})] \tag{8.19}$$

Degrees of Freedom (*df*)

The (*df*) for individual factor is one less than the number of levels, and (*df*) for interaction A and B are the product of (*df*) of factor A and B. If *n* is equal to the number of readings, and α, β are levels for factors A and B, respectively, then *df* for single observation are as follows:

$$df\,(factor.A) = \alpha - 1$$
$$df\,(factor.B) = \beta - 1$$
$$df\,(Interaction\,A \times B) = (\alpha - 1)(\beta - 1) = \alpha\beta - \beta - \alpha + 1$$
$$df\,(total) = n - 1 = df\,(factor.A) + df\,(factor.B) + df\,(Interaction.A \times B)$$
$$= [(\alpha - 1) + (\beta - 1) + (\alpha\beta - \beta - \alpha + 1)] = \alpha\beta - 1$$

For replicated observation design, the (*df*)s are different for "total" and "error":

$$df\,(total) = n - 1 = \alpha\beta r - 1$$
$$df\,(error) = n - 1 = df\,(total) - [df\,(factor.A) + df\,(factor.B)$$
$$+ df\,(Interaction.AB)]$$
$$= (\alpha\beta r - 1) - [(\alpha - 1) + (\beta - 1) + (\alpha\beta - \beta - \alpha + 1)] = \alpha\beta(r - 1)$$

Testing for Hypothesis

a. Factor A:
   $H_{0,A}$: Factor A is not significant ($\mu_1 = \mu_2 = \mu_3$)
   $H_{a,A}$: Factor A is significant (not all means are equal)
   Reject $H_{0,A}$ if $F_A(calculated) > F_{\alpha',(a-1),\alpha\beta(r-1)}$, with significance level (α') value
   where F_A is calculated and $F_{\alpha',(a-1),\alpha\beta(r-1)}$ is obtained from the F-Table in the Appendix. (Note that here α' is different from α.)
b. Factor B:
   $H_{0,B}$: Factor B is not significant ($\mu_1 = \mu_2 = \mu_3$)
   $H_{a,B}$: Factor B is significant (not all means are equal)
   Reject $H_{0,B}$ if $F_B(calculated) > F_{\alpha',(\beta-1),\alpha\beta(r-1)}$
   where F_B is calculated and $F_{\alpha',(\beta-1),\alpha\beta(r-1)}$ is obtained from the table in the Appendix.
c. Interaction AB:
   $H_{0,AB}$: No significant interaction between A and B
   $H_{a,AB}$: Significant interaction between A and B
   Reject $H_{0,AB}$ if $F_{AB}(calculated) > F_{\alpha',(a-1)(\beta-1),\alpha\beta(r-1)}$
   where F_{AB} is calculated and $F_{\alpha',(a-1)(\beta-1),\alpha\beta(r-1)}$ is obtained from the F-Table in the Appendix.

EXAMPLE 8.5.   A manufacturing company is interested in the number of defective parts produced by its operators. A quality/manufacturing engineer was assigned for this project. She proposed three types of procedures to be followed by the operators. To find out if there was a difference in the three types of procedures, an experiment was conducted with Factor 1 factor having three levels:

1. Less than two years work background
2. Two to five years work background
3. More than five years work background

The second factor was a systematic procedure, one level for each one of the three recommended procedures. The following data show the average number of non-defective parts produced per day over a one-week period for each operator.

Proposed Procedure Table

	Procedure (factor B)		
Background (Factor A)	1	2	3
Less than 2 Years	22.5, 25.7, 20.1	19.2, 18.6, 17.3	23.4, 24.1, 21.3
Between 2 and 5 Years	23.5, 24.0, 25.2	19.0, 15.2, 14.5	22.1, 21.4, 24.5
More than 5 Years	27.5, 29.6, 30.0	26.4, 27.5, 24.6	26.7, 28.5, 27.0

Determine if there is any difference in the outcome of three proposed procedures using 10% significant level ($\alpha' = 0.10$).

Let A = background and B = procedure.

Guideline Table

Background	1	2	3	Total
Less than 2 Years	22.5 + 25.7 + 20.1 = 68.3	19.2 + 18.6 + 17.3 = 55.1	23.4 + 24.1 + 21.3 = 68.8	$T_1 = 192.2$
Between 2 and 5 Years	23.5 + 24.0 + 25.2 = 72.7	19.0 + 15.2 + 14.5 = 48.7	22.1 + 21.4 + 24.5 = 68.0	$T_2 = 189.4$
More than 5 Years	27.5 + 29.6 + 30.0 = 87.1	26.4 + 27.5 + 24.6 = 78.5	26.7 + 28.5 + 27.0 = 82.2	$T_3 = 247.8$
Total	$S_1 = 228.1$	$S_2 = 182.3$	$S_3 = 219$	$T = 629.4$

Sums of the squares are

Factor A (background): $SS(factor.A) = \dfrac{1}{\beta r}\left(T_1^2 + T_2^2 + \cdots + T_m^2\right) - \dfrac{T^2}{\alpha\beta r}$

$SS(factor.A) = \dfrac{1}{3 \times 3}\left(192.2^2 + 189.4^2 + 247.8^2\right) - \dfrac{629.4^2}{3 \times 3 \times 3} = 241.10$

where T = total of all $n = \alpha\beta r$ readings—that is, $T = T_1 + T_2 + \cdots + T_m.$

Factor B (guideline): $SS(factor.B) = \dfrac{1}{\alpha r}\left(S_1^2 + S_2^2 + \cdots + S_k^2\right) - \dfrac{T^2}{\alpha\beta r}$

$SS(factor.B) = \dfrac{1}{3 \times 3}\left(228.1^2 + 182.3^2 + 219^2\right) - \dfrac{629.4^2}{3 \times 3 \times 3}$

$= 14,802.66 - 14,672.01 = 130.64$

Interaction AB:

$SS(Interaction.AB) = \dfrac{\sum R^2}{r} - \left[SS(factor.A) + SS(factor.B) + \dfrac{T^2}{\alpha\beta r}\right]$

where ΣR^2 is the sum of all the squares of replicate totals.

$\sum R^2 = 68.3^2 + 55.1^2 + 68.8^2 + 72.7^2 + 48.7^2 + 68.0^2 + 87.1^2 + 78.5^2 + 82.2^2$

$= 45,220.82$

$SS(Interaction.AB) = \dfrac{45,220.82}{3} - \left[241.10 + 130.64 + \dfrac{629.4^2}{3 \times 3 \times 3}\right]$

$= 15,073.61 - 15043.75 = 29.86$

Total: $SS(total\ replicate) = \sum x^2 - \dfrac{T^2}{\alpha\beta r}$

$SS(total) = \left(22.2^2 + 25.7^2 + 20.1^2 + \cdots + 26.7^2 + 28.5^2 + 27.0^2\right) - \dfrac{692.4^2}{3 \times 3 \times 3}$

$= 15,123.82 - 14,672.01$

$= 451.81$

$SS(error) = SS(total\ replicates) - [SS(factor\ A) + SS(factor\ B)$

$+ SS(Interaction\ AB)]$

$= 451.81 - (241.10 + 130.64 + 29.86)$

$= 50.21$

$MS(factor.A) = \dfrac{SS(factor.A)}{\alpha - 1} = \dfrac{241.10}{3-1} = 120.55$

$MS(factor.B) = \dfrac{SS(factor.B)}{\beta - 1} = \dfrac{130.64}{3-1} = 65.32$

$$MS(Interaction.AB) = \frac{SS(Interaction.AB)}{(\alpha-1)(\beta-1)} = \frac{29.86}{(3-1)(3-1)} = 7.465$$

$$MS(error) = \frac{SS(error)}{\alpha\beta(r-1)} = \frac{50.21}{3\times3(3-1)} = 2.79$$

$$F_A = \frac{MS(factor.A)}{MS(error)} = \frac{120.55}{2.79} = 43.20$$

$$F_B = \frac{MS(factor.B)}{MS(error)} = \frac{65.32}{2.79} = 23.41$$

$$F_{AB} = \frac{MS(Interaction.AB)}{MS(error)} = \frac{7.465}{2.79} = 2.68$$

Two-Factor ANOVA Table

Source of inconsistency (variation)	Degrees of freedom (df)	Sum of squares (SS)	Mean squares (MS)	Statistical test (F)
Factor A	$\alpha - 1 =$ $3 - 1 = 2$	SS (factor A) $= 241.10$	$MS(factor.A) =$ $\frac{SS(factor.A)}{\alpha-1}$ $= 120.55$	$F_A =$ $\frac{MS(factor.A)}{MS(error)}$ $= 43.20$
Factor B	$\beta - 1 =$ $3 - 1 = 2$	SS (factor B) $= 130.64$	$MS(factor.B) =$ $\frac{SS(factor.B)}{\beta-1}$ $= 65.32$	$F_B =$ $\frac{MS(factor.B)}{MS(error)}$ $= 23.41$
Interaction A × B	$(\alpha-1)(\beta-1) =$ $2 \times 2 = 4$	SS (Interaction AB) $= 29.86$	$MS(Interaction.AB)$ $= \frac{SS(Interaction.AB)}{(\alpha-1)(\beta-1)}$ $= 7.47$	$F_{AB} =$ $\frac{MS(Interaction.AB)}{MS(error)}$ $= 2.68$
Error	$\alpha\beta(r-1) =$ $3 \times 3(3-1)$ $= 18$	SS (error) $= 50.21$	$MS(error) =$ $\frac{SS(error)}{\alpha\beta(r-1)}$ $= 260.92$	
Total	$\alpha\beta r - 1 =$ $(3 \times 3 \times 3) - 1$ $= 26$	SS (total) $= 451.81$		

So make a conclusion.

a. Factor A:

$H_{0,A}$: Factor A is not significant ($\mu_1 = \mu_2 = \mu_3$)

$H_{a,A}$: Factor A is significant (not all means are equal)

Reject $H_{0,A}$ if $F_A > F_{\alpha,(a-1),\alpha\beta(r-1)}$

where F_A is calculated and $F_{\alpha,(a-1),\alpha\beta(r-1)}$ is obtained from the F-Table in the Appendix. Thus,

$$F_{0.10,(3-1),3\times3(3-1)} = F_{0.10,2,18} = 2.62$$

$43.20 > F_{0.10,2,18} = 2.62$, conclude that there is a significant difference in the three guidelines.

b. Factor B:

$H_{0,B}$: Factor B is not significant ($\mu_1 = \mu_2 = \mu_3$)

$H_{a,B}$: Factor B is significant (not all means are equal)

Reject $H_{0,B}$ if $F_B > F_{\alpha,(\beta-1),\alpha\beta(r-1)}$

where F_B is calculated and $F_{\alpha,(\beta-1),\alpha\beta(r-1)}$ is obtained from the F-Table in the Appendix. Thus,

$$F_{0.10,(3-1),3\times3(3-1)} = F_{0.10,2,18} = 2.62$$

$23.41 > F_{0.10,2,18} = 2.62$, conclude that there is a significant difference in the three guidelines.

c. Interaction AB:

$H_{0,AB}$: No significant interaction between A and B

$H_{a,AB}$: Significant interaction between A and B

Reject $H_{0,AB}$ if $F_{AB} > F_{\alpha,(a-1)(\beta-1),\alpha\beta(r-1)}$

where F_{AB} is calculated and $F_{\alpha,(a-1)(\beta-1),\alpha\beta(r-1)}$ is obtained from the F-Table in the Appendix. Thus,

$$F_{0.10,(3-1)(3-1),3\times3(3-1)} = F_{0.10,4,18} = 2.29$$
$$2.68 > F_{0.10,4,18} = 2.29, \text{ interaction is significant.}$$

Another hypothesis test is called a *z*-test using one population proportion.

z-Test Using One Population Proportion (for Larger Samples $np_o > 5$ and $n(1 - p_o) > 5$)

The steps are as follows:

1. Set hypothesis test conditions for null test and alternative
   For two-tailed test: p = mean
   $H_o: p = p_o$
   $H_a: p \neq p_o$
   Verify the values of $np_o > 5$ and $n(1 - p_o) > 5$.

Reject H_o if absolute value of calculated z is larger than $z_{\alpha/2}$ from the z-Table in the Appendix. ($|z| > z_{\alpha/2}$), where

$$z = (\hat{p} - p_o)\left(\frac{p_o(1 - p_o)}{n}\right)^{-\frac{1}{2}}$$

$$= (\text{estimate of } p - \text{hypothesized value})(\text{estimated standard deviation of } p)$$

$$(8.20)$$

where
$\hat{p}$ = estimate of p ($\hat{p}$ is normal random variable with mean equal to p)
   or sample proportion having a specified attribute.

$$\hat{p} = \frac{x}{n}$$

x = defect readings
p_o = hypothesized value
n = sample size

2. For one-tailed test:
   $H_o: p \le p_o$ left-tailed $H_o: p \ge p_o$ right-tailed test
   $H_a: p > p_o$ $H_a: p < p_o$
   Reject H_o if $z_{calculated} > z_\alpha$ Reject H_o if $z_{calculated} < -z_\alpha$
   where z is calculated and z_α is obtained from the z-Table.

EXAMPLE 8.6.   In a manufacturing company a quality engineer was assigned to inspect 200 randomly selected parts from a lot. He found 15 defective parts. He thinks that a good defective amount is about 3% ($p_o = 0.03$). Can he conclude that the actual proportion of defective calculators is anything other than 5%? Use a significance level of $\alpha = 0.05$.

1. The inspector uses a two-tailed test to find out if p is different from 5%.

$$H_o: p = 0.03$$
$$H_a: p \ne 0.03$$

2. $np_o = (200)(0.03) = 6 > 5$ and $n(1 - p_o) = (200)(1 - 0.03) = 194 > 5$. So both are greater than 5. Then, z-test applies:

$$z = (\hat{p} - p_o)\left(\frac{p_o(1 - p_o)}{n}\right)^{-\frac{1}{2}} = (\hat{p} - 0.03)\left(\frac{0.03(1 - 0.03)}{200}\right)^{-\frac{1}{2}}$$

3. At $\alpha = 0.05$ the null H_o and alternative H_a test will be

Reject H_o if $|z| > 1.96$, (use z-Table at $z_{\alpha/2} = z_{0.025} = 1.96$. Note that at area = $0.50 - 0.025 = 0.475$ (Figure 8.7), the z-value from the z-Table equal to 1.96).

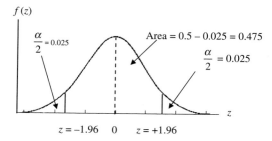

Figure 8.7   Normal distribution curve for the Example 8.6.

4. Calculate z using estimate $\hat{p}$ (p hat)

$$\hat{p} = \frac{x}{n} = \frac{15}{200} = 0.075$$

$$z = (\hat{p} - p_o)\left(\frac{p_o(1 - p_o)}{n}\right)^{-\frac{1}{2}}$$

$$= (0.075 - 0.03)\left(\frac{0.03(1 - 0.03)}{200}\right)^{-\frac{1}{2}}$$

$$= 3.73$$

Since $|z| = |3.73| = 3.73 > z_{\alpha/2} = z_{0.05/2} = 1.96$. So H_o is rejected.
5. Based on step 4, the company is not meeting its rejection parts (or defective parts) percentage.

χ^2 Chi-Square Test (Hypothesis Test) for Population Standard Deviation

The conditions and procedures are given Example 8.7.

EXAMPLE 8.7.   A dimension of a product in the manufacturing process has a standard deviation of $\sigma_o = 0.035$. In order to improve the process, the manufacturing engineer, after making a change, obtained 25 samples and measured them. The standard deviation was found to be $s = 0.038$. The old data showed the population is normally distributed. Use $\alpha = 0.05$ significance level.

1. Condition: population is normally distributed.
2. Apply α value significance level.

3. Set hypothesis test $H_o: \sigma = \sigma_o$ and $H_o: \sigma \neq \sigma_o$ (two tails) or one-tailed tests $H_o: \sigma \leq \sigma_o$ and $H_a: \sigma > \sigma_o$ or $H_o: \sigma \geq \sigma_o$ and $\sigma < \sigma_o$, where here $\sigma > 0.035$ right tail (note that the hypothesis for $H_o: \sigma \leq \sigma_o$ and $H_a: \sigma > \sigma_o$ are the same if one writes H_o or H_a in terms of σ or σ^2 the testing steps are the same for either condition).

4. Obtain critical values from the χ^2-Table in the Appendix just like the t-Table using degrees of freedom $= n - 1$.
   a. Two-tail test: values are $\chi^2_{1-\alpha/2}$ and $\chi^2_{\alpha/2}$
   b. Left-tail test: value is $\chi^2_{1-\alpha}$
   c. Right-tail test: value is χ^2_{α}
   Here, the critical value for the right tail from the χ^2-Table (critical values of χ^2) at $\chi^2_{0.05}$ and $df = n - 1 = 25 - 1 = 24$ is $\chi^2 = 36.415$.

5. Determine test statistic χ^2 using formula (8.21)

$$\chi^2 = \frac{s^2(n-1)}{\sigma_o^2} \tag{8.21}$$

where s is a sample standard deviation comparing to σ_0 value

$$\chi^2 = \frac{s^2(n-1)}{\sigma_o^2} = \frac{(0.038)^2(25-1)}{(0.035)^2} = 28.29$$

6. Make a conclusion.
   a. For a two-tailed condition:

$$H_o: \sigma = \sigma_o$$
$$H_a: \sigma \neq \sigma_o$$

Test statistic: $\chi^2 = \dfrac{s^2(n-1)}{\sigma_o^2}$

Reject H_o if $\chi^2 > \chi^2_{\alpha/2,n-1}$ or $\chi^2 > \chi^2_{1-\alpha/2,n-1}$

   b. One-tailed test:

$H_o: \sigma \leq \sigma_o$	$H_o: \sigma \geq \sigma_o$
$H_a: \sigma > \sigma_o$	$H_a: \sigma < \sigma_o$
Reject H_o if $\chi^2 > \chi^2_{\alpha,n-1}$	Reject H_o if $\chi^2 > \chi^2_{1-\alpha,n-1}$

The test procedure says to reject if the χ^2 test statistic value is in the reject region, but do not reject otherwise.

Here, the value of preceding $\chi^2 = 36.415$ from Appendix χ^2-Table is in rejection of H_o. Since 28.29 is not in the rejection region (28.29 < 36.415), do not reject H_o.

7. In conclusion, at $\alpha = 0.05$, data do not show that population standard deviation has increased.

Process Capability Ratio for Short Term and Long Term

Capability index is the repeatability of a production process based on the customer specification limit (LSL and USL). The capability index C_p does not consider how the data is centered. On the other hand, C_{pk} is used to distinguish the actual tolerance with specified limits. If the process is in statistical process control and the process data are centered on the target so that the bell-shaped curve is symmetrical, then the capability index C_p can be determined using Equation 8.22.

$$C_p = \frac{USL - LSL}{6\sigma} = \frac{specification/width}{process/width} \tag{8.22}$$

The general rule of interpreting C_p is

$C_p < 1.0$ inadequate; process variation is higher than specification and has more defects.

$1.0 \leq C_p < 1.33$ adequate; the process is acceptable and just meets specifications; 0.27% defects will be produced for short term, and the process is still required to be centered. Average industry quality falls in this category.

$C_p \geq 1.33$ good; the process variation is less than specification, but possible defects and process are required to be centered or to be maintained in control.

Figures 8.8 and 8.9 illustrate the preceding concepts.

Here, C_p does not take the mean into consideration. Thus, it does not address how well the process mean is centered to the target value. However, C_{pk} does take the mean into account. Therefore, C_{pk} is the more realistic capability index.

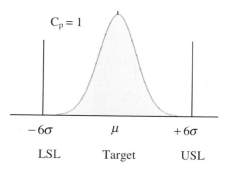

Figure 8.8 Standard normal distribution curve centered at the target.

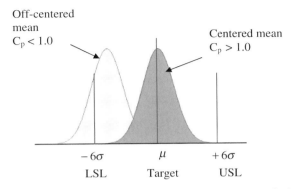

Figure 8.9   Standard normal distribution curve (shifted 1.5 sigma) for long term.

C_{pk} value is taken as the smallest of either Equation 8.23 or 8.24 or $C_{pk} =$ min [C_{pk}(USL), C_{pk}(LSL)]. Also see Equations 8.18 and 8.19. $P_{pk} =$ min [P_{pk}(USL), P_{pk}(LSL)]. Equation 8.23 illustrates C_{pk} using the upper specification limit:

$$C_{pk}(USL) = \frac{USL - \mu}{3\sigma} \tag{8.23}$$

And Equation 8.24 represents C_{pk} using the lower specification limit:

$$C_{pk}(LSL) = \frac{\mu - LSL}{3\sigma} \tag{8.24}$$

Rearranging Equation 3.14 concludes Equation 8.25

$$\sigma = \frac{USL - \mu}{z} \tag{8.25}$$

then, by substituting Equation 8.25 in Equation 8.23, the Equation 8.26 results in

$$C_{pk} = \frac{USL - \mu}{3\left(\dfrac{USL - \mu}{z}\right)} = \frac{z_{USL}}{3} \tag{8.26}$$

The C_{pk} Equation 8.27 gives us information on how far the distance of the actual mean is from the process target.

$$C_{pk} = C_p(1 - K) \tag{8.27}$$

where $K = \dfrac{|T - \mu|}{\dfrac{USL - LSL}{2}} = \dfrac{2(|T - \mu|)}{USL - LSL}$

or replacing the K value in Equation 8.27 C_{pk} becomes Equation 8.28.

$$C_{pk} = C_p\left(1 - \frac{2(|T - \mu|)}{USL - LSL}\right) \tag{8.28}$$

When the target equals mean $(T = \mu)$ in Equation 8.28, then Equation 8.29 is concluded. In other words, if the process mean remains centered on the target nominal, then the value of $K = 0$.

$$C_{pk} = C_p \tag{8.29}$$

where T is the target and μ is the actual mean.

Tables 8.4 and 8.5 show the nonconformance quantity for different C_{pk} values when the process is centered at the target and shifted by 1.5 sigma from the target.

Tables 8.4 and 8.5 were constructed using Equations 8.22 through 8.26. The graphical representation of Table 8.5 is shown in Figure 8.10.

Thus, as the sigma level increases, product quality and profitability also increase. This means that cycle time/Takt time reduces, quality checks are minimized, operating cost goes down, customer satisfaction goes up, and 80% of lead time is reduced, which is caused by 20% workstation time traps.

Table 8.4

Impact of Short-Term Process Capability When the Process Is Centered on the Target

C_{pk}	Defects per million	Sigma level
0.33	317,310	1.00
0.50	133,614	1.50
0.67	45,500	2.00
0.75	24,448	2.25
1.000	2,700	3.00 (Traditional quality)
1.167	465	3.50
1.333	63	4.00
1.500	6.8	4.50
1.667	0.6	5.00
2.000	0.002	6.00

Table 8.5

Impact of Long-Term Process Capability When the Process Is Shifted by 1.5 Sigma from the Target

C_{pk}	Defects per million (expected nonconformance)	Sigma level
0.00	500,000	1.5 Sigma
0.17	308,300	2.0 Sigma
0.33	158,650	2.5 Sigma
0.50	66,807	3.0 Sigma (Traditional quality)
0.67	22,700	3.5 Sigma
0.83	6,220	4.0 Sigma
1.00	1,350	4.5 Sigma
1.17	233	5.0 Sigma
1.33	32	5.5 Sigma
1.50	3.4	6.0 Sigma

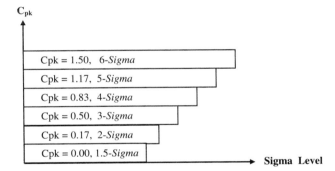

Figure 8.10   Graphical representation of Table 8.4, sigma level increases as C_{pk} increases.

Differences between C_{pk} and P_{pk}

Process Capability (C_p): Represents process capability when the process is centered at the target (short term with no sigma shift).

Process Capability Index (C_{pk}): Represents process capability when the process is shifted by 1.5 sigma (long term). C_{pk} uses an estimated population standard deviation.

$$\sigma = \sqrt{\frac{\sum_{i=1}^{n}(x_i - \mu)^2}{N}}$$

Process Performance (P_p): Represents process performance when the process is centered at the target (short term with no sigma shift).

Process Performance Index (P_{pk}): Represents process performance when the process is shifted by 1.5 sigma (long term). P_{pk} uses actual sample standard deviation.

$$S = \sqrt{\frac{\sum_{i=1}^{n}(x_i - \bar{x})^2}{(n-1)}}$$

Recall C_{pk} from the previous discussion $C_{pk} = \min\,[C_{pk}(USL), C_{pk}(LSL)]$, $P_{pk} = \min[P_{pk}(USL), P_{pk}(LSL)]$.

As shown before, C_{pk} using the upper and lower specification limits:

$$C_{pk}(USL) = \frac{USL - \mu}{3\sigma}, \; C_{pk}(LSL) = \frac{\mu - LSL}{3\sigma}$$

where estimated

$$\sigma = \sqrt{\frac{\sum_{i=1}^{N}(x_i - \mu)^2}{N}}$$

Similarly, process performance for the long term:

$$P_{pk} = \frac{USL - \mu}{3S} \tag{8.30}$$

where

$$S = \sqrt{\frac{\sum_{i=1}^{n}(x_i - \bar{x})^2}{(n-1)}}$$

and Equation 8.31 represents P_{pk} using the lower specification limit:

$$P_{pk} = \frac{\mu - LSL}{3S} \tag{8.31}$$

Identification of Customer Specification Range

As previously discussed, customer satisfaction is a function of quality, cost, and shipping speed, as shown in the following model:

Customer satisfaction $= f$ (*Quality, Cost, Shipping speed*)

Prior to design layout, the design engineering team should understand customer specification limits (critical to quality) through VOC. A comparison of the current and future components with customer requirements must be outlined. Then designers should apply the design for Lean *Six Sigma* (DFLSS) capability at the early stages (review the *Six Sigma* ergonomic in Chapter 5). In addition, production as well as manufacturing engineers should review the design for all the possible issues before the tooling is completed.

Identification of Manipulated (Cause) and Response Variables

The process engineer should identify independent variables (experimental factors) that affect dependent (response) variables in improving and contributing to the mean of normal distribution.

8.3 EVALUATION OF EXISTING PROCESS SIGMA/ BASELINE SIGMA

Compute sigma and z-value using the following parameters:

a. Natural tolerance of a process.
b. Determine defect-free probability (DFP) or defect per million opportunities (DPMO) for the existing data using DPMO = [(Total Defects)/(Total Opportunities)] $\times 10^6$.
c. Obtain and understand the following process variation information from the previous sections. The parameters are μ, σ, C_p, C_{pk}, P_p, P_{pk}, gage R&R, Pareto diagram, histogram, and control charts.
d. Determine the z-value and sigma (σ) level of the standard deviation of statistical population using any statistical software or by using Microsoft Excel function "= normsdist (z) = normsinv (probability)" (see Equation 2.1).
e. Determine the sigma shift.
f. Calculate the process sigma rating from the defect-free probability.
g. Define a process baseline performance for customer quality.

8.4 DATA ANALYSIS

8.4.1 PHASE II: PROCESS ANALYSIS

Most dominant factors (parameters) and source of variations that contribute to sigma mean and variance will be recognized. Ergonomic factors will be examined in the process. Thus, thorough analysis of the existing process is required for redesign in some cases or future designs of process/product. This section will describe the following goals of Lean *Six Sigma.*

Process Analysis Objectives

Phase II Lean *Six Sigma* goals, objectives, and metrics are outlined as follows:

a. Business process and data analysis.
b. Root cause and effect analysis.
   Identify and list the source of variations and root causes of a few dominant factors. Refer to Equation 6.3 for process performance function in manufacturing and production for regression modeling.
c. Create multi-vari analysis that helps to identify the process performance variables.
d. Create benchmark performance metrics/evaluate current process against best practices in the manufacturing organizations.
e. Determine variance components and sources by identifying the process factors (most dominant X's), process delay factors, and estimating process capability such as hypothesis testing, p-value, and so on.
f. Analyze the $\Delta\sigma$ (difference of existing sigma and objective/goal sigma levels).
g. Assess correlation.
h. Analyze the process for value-added and nonvalue-added factors. (Review Chapter 6.)
i. Implement experimental design strategies to analyze the problems.
j. Apply *Six Sigma* ergonomics.

Analysis Tools and Techniques

1. Cause-and Effect-Matrix
   Cause-and-effect matrices are used to study a problem or improvement opportunity in identifying root causes. They focus on causes, not symptoms.

The cause-and-effect diagrams (also called fishbone) are used to explore all the potential or real causes (or inputs) that result in a single effect (or output). Causes are laid out based on their level of importance or detail. This can help one search for root causes and identify areas where there may be problems. An example of cause-and-effect diagrams for injection molding processes, as shown

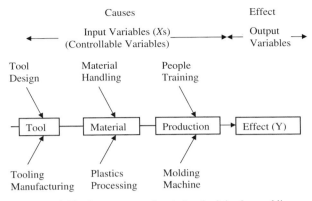

Figure 8.11 Process map of variation for injection molding.

in Figure 8.11, is three-section tooling, material, and production. Each section is caused by different categories of variations.

Establish Cause and Effects

Ask *why* five times. Identify root causes by analyzing potential causes as long as one can ask *why* and get an answer: that potential cause was not the root cause. For example, ask the "five *whys*" questions—in this case, plastics injection molding of a large tube-shaped part had uneven wall thickness.

1. Why was the wall thickness uneven?
   It was due to sink marks on the last filling profile of the part.
2. Why did sink marks appear in the parts?
   It was due to lack of polymer melt in the wall thickness.
3. Why was there a lack of polymer material in the wall thickness?
   It was due to runners freezing too fast.
4. Why were runners freezing too fast?
   It was due to material temperatures being too low.
5. Why were material temperatures too low?
   It was due to bad thermocouples.

Normally, the third question will resolve the majority of the problems.

2. Multi-Vari Cause Charts and Analysis

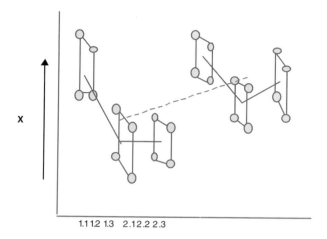

1.1 1.2 1.3   2.1 2.2 2.3

It is used to identify and reduce possible independent variables (*X*s) or group of causes of variation to a much smaller group of control factors. Thus, constructed examples to display the changes for three major types of variations are as follows:

a. Positional variation
   i. Within the units or unit family
   ii. Variations within a single unit—for example, left side versus right side or top versus bottom
   iii. Variations across a single unit—for example, thickness of a printed circuit board
   iv. Variations from plant cell to plant cell, machine to machine, operator to operator, mold to mold, and cavity to cavity
b. Cyclical variation
   i. Unit-to-unit family
   ii. Variation between consecutive units in the same time frame
   iii. Batch-to-batch variation
   iv. Lot-to-lot variation
c. Temporal variation
   i. Time to time family
   ii. Month to month
   iii. Week to week
   iv. Shift to shift
   v. Day to day
   vi. Hour to hour

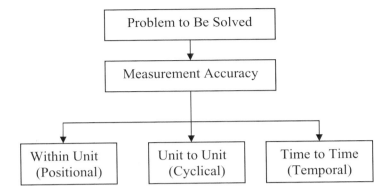

3. Scatter diagram and box plot
   a. Scatter diagram (Pearson's R value) illustrates the visual view of the qualitative relationship, either linear or nonlinear, that exists between two variables (of input and output) using an X and Y diagram. Furthermore:
      i. It provides both a visual and statistical means to test the strength of a potential relationship.
      ii. It supplies the data to confirm that two variables are related.
      iii. Steps are
         1. Collect paired data samples.
         2. Draw the x- and y-axis of the graph.
         3. Plot the data.
         4. Interpret the outcome of the plot, such that values can be only between (−1, +1). For example,
            a. Strong positive linear slope in x-y coordinate means "perfect (strong) positive correlation," as shown here.

$$Slope = \frac{\Delta y}{\Delta x} = +1.0.$$

(Variable Y)

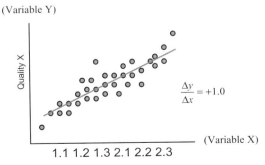

$\frac{\Delta y}{\Delta x} = +1.0$

(Variable X)

1.1 1.2 1.3 2.1 2.2 2.3

b. Strong negative linear slope in x-y coordinate means "perfect (strong) negative correlation," as shown here:

$$Slope = \frac{\Delta y}{\Delta x} = -1.0.$$

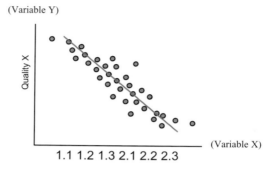

(Variable Y)

Quality X

1.1 1.2 1.3 2.1 2.2 2.3   (Variable X)

c. Scatter (nonlinear) data all over the x-y coordinate means "no correlation"—for example, $Slope = \frac{\Delta y}{\Delta x} = 0.0.$

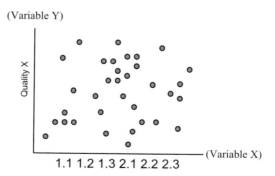

(Variable Y)

Quality X

1.1 1.2 1.3 2.1 2.2 2.3   (Variable X)

Pearson Correlation
Pearson is another method of distinguishing the different levels of correlation.

Pearson Correlation

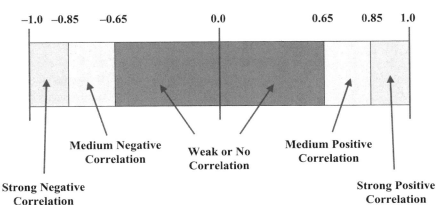

b. Box and whisker plot (box plot)

A box plot is used to show the distribution of a single variable in a data set. It is a graphical representation of data using five measures:

1. Median
2. First quartile: 25% of values are less than Q_1, and 75% are larger than Q_1. Q_2 divides data into two equal parts—hence Q_2 = median.
3. Third quartile: 75% of values are less than Q_3, and 25% are greater than Q_3. The difference between Q_3 and Q_1 is called the interquartile range (IQR).
4. Smallest value
5. Largest value

This helps to visualize the center, the spread, and the skewness of a data spread, as follows:

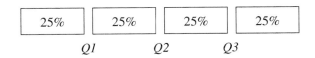

4. Identify the sources of variations and root causes (independent variables) of a few dominant problems.

   i. Develop a single/multiple regression relationship, $Y = f(x) = ax + b$
   ii. Keep in mind the process performance model from Chapter 6, Equation 6.3: *Performance* = f [*Design* (*product and tool*), *Materials, Tooling, Process Conditions*]

      iii. Review inadequate design windows.

      iv. Examine for unstable material properties.

      v. Distinguish inadequate tooling capability.

      vi. Test for lack of process capability.

5. Create benchmarking performance metrics.
6. Use a value-added analysis approach:
   i. Draw process flowchart.
   ii. Identify the activities that added value.
   iii. Identify the nonvalue-added activities.
   iv. Decide whether activities should be kept, combined, or eliminated.
   v. Determine appropriate actions.
7. Identify process capability variables. Capability index is the repeatability of the production process based on the customer specification limits (USL & LSL).
8. Apply the following mathematical and statistical analysis as necessary: z, C_p, C_{pk}, P_p, P_{pk}, DPMO, PPM, correlation, regression (single/multiple), and hypothesis testing (continuous and discrete), where it applies based on the importance problem.
9. Analyze failure factors.
10. Analyze the $\Delta\sigma$—that is, $6\sigma - 3.5\sigma$ (current σ) $= 2.5\sigma$.

8.5 OPTIMIZATION AND IMPROVEMENT

8.5.1 PHASE III: PROCESS IMPROVEMENT

Optimization and Improvement Objectives

a. Develop and evaluate solutions.
   - Develop a concept prototype (technology), if possible.
   - Optimize the concept design.
   - Test capability verification.
b. Implement variation reduction.
c. Standardize the process.
d. Assess risk factors (performance verification).
e. Apply the laws of Lean *Six Sigma*.
f. Design or redesign for Lean *Six Sigma*.
   i. Conduct a benchmark (distinguishing of best practices) study.
   ii. Benchmark the following processes:
      1. Strategic point of view.
         - Identify the best in class companies that use the methods and technologies in their practices.

- Continuously improve your product or services to be the best compared with the best companies in the market.
- Integrate competitive ideas, performance, and information into the creation of business goals and objectives on all levels.
  2. Administration, finance, marketing/sales, design, tooling, material/process, manufacturing/production, software, and shipping
g. Create variable relations.
h. Reduce variables that result in nonconformances in existing processes.
i. Fit design to material, machine, tool, and human capabilities.
j. Improve new opportunities that will close gaps.

Improvement Tools and Techniques

- **Experimental Design (DOE robust design).**

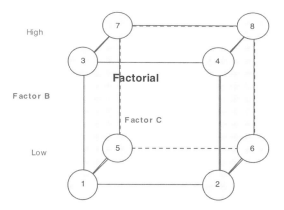

- An efficient technique of experimentation that identifies key process input variables (Xs) and their optimum settings that affect process mean and output variations (Ys) with minimum testing.
- Establish critical and noncritical variable selection.
- Identify dominant process performance-improving variables.
- Introduce and verify the selected variables that contribute to mean improvement.
- Apply linear and nonlinear multivariable regression modeling and analysis.
- Apply experimental design in improving the problem.

- Establish a mathematical model.
- Evaluate performance improvement.
- Eliminate process steps that are unnecessary. Convert the steps to those that will achieve Lean *Six Sigma*. Use the statistical tool to optimize your process, such as the following:
  - DOE, full factorial design, fractional factorial design, response surface analysis, Taguchi, and ANOVA.
  - Establish a process model (see case studies in Chapters 9 and 10).
  - Optimize the process model through simulation and other techniques.
- Implement the optimized process in the production line.
- **Response Surface Techniques** (nonlinear analysis techniques); see Chapter 10 for details.

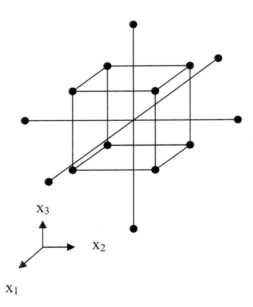

The purpose of Response Surface Technique modeling is to obtain maximum or minimum response $y = f(x)$ at an optimum value (see Chapter 10 for more details on the response surface analysis modeling). Response Surface Techniques use the steepest ascent simulation—for example, variables are changed in the improvement of response until the response increases or decreases.

- **Deployment Flowchart (Process Mapping)**

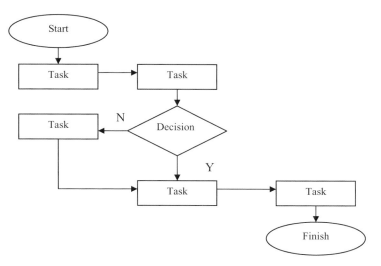

The above flowchart is an example of a process path from start to finish.

• **Tree Diagram**

A tree diagram breaks any broad goal into levels of detailed action plans. It also visually displays connectivity between goals and action plans.

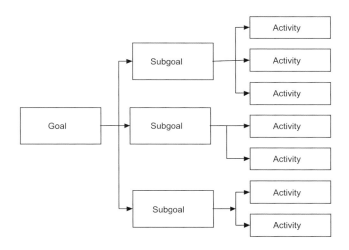

Failure Mode Effects Analysis (FMEA)

In the 1960s FMEA was first formally used during the Apollo space program. In the late 1960s FMEA became a key tool in the chemical process industries as a way to improve safety. Then, in 1974 the U.S. Navy introduced Mil-Std-1629 regarding the use of FMEA. In the late 1970s the automotive industries adapted the FMEA techniques, initially for safety improvement, and later for product liability reduction. In the mid-1990s QS-9000 QMS mandated the formal use of FMEA and continued to *Six Sigma*, Lean program, manufacturing industries in designing product/process.

FMEA is an analytical approach for preventing defects by prioritizing potential problems and their resolution. Further, FMEA is a systemized group of activities intended to recognize and evaluate the potential failure of a product/process and its effects. In addition, it identifies actions that could eliminate or reduce the probability or severity of the potential failure and effects (document the process). Finally, it is a proactive quality planning and improvement process, associated with risk analysis and the calculation of reducing risks. These are the FMEA process steps:

1. Identification of failure modes
2. Determination of the effects of failure modes
3. Implementation of contingency action plans

FMEA is continuous in risk analysis and identifies the ways in which a product, process, or service can fail. It estimates the risks associated with specific causes. Then it takes into consideration the severity of product failure, which may range without warning from hazardous to moderate to none. Eventually, it prioritizes the actions that should be taken to reduce the risks. Where do risks come from? The risk factors are shown here:

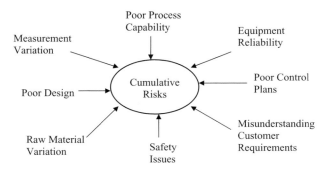

Managing process and product risks are one of the important aspects of this subject. This involves the probability and reliability, as discussed in previous chapters. In brief, from a management point of view, risk management is equal to identification, quantification, response, and lessons learned. The three-stage process includes the FMEA and risk prioritization number (RPN).

Stage 1. Identification of Potential Failure Mode and the Effect
To identify FMEA, one can use a block diagram of subsystem or subprocess activities, a tree diagram to capture linkage and connectivity logic, brainstorm techniques to collect areas of potential failure modes, and data to capture areas of actual failure mode.

Stage 2. Find the Modes and Effects of Failures
Knowing that failures cause defects, one may ask, what are defects? Products and services are defective because they deviate from the intended condition from the provider. They are unsafe due to design defects even though they meet specifications or they are incapable of meeting their claimed level of performance. Hence, they are dangerous because they lack adequate warning and instructions.

Where do you find such failure modes? One may search in the area of manufacturing operation for opportunities as product deformed, cracked, leaking, misaligned, broken, eccentric, loose, jams, sticking, imbalance, short circuit, or wear and tear. The typical failure effects may come from process noise, erratic machine stops, poor appearance, unstable operation, inoperative equipment, and rough processes.

A sample of a Failure Mode Effects Analysis (FMEA) table is given here. (See Chapter 9 for a detailed FMEA table.)

Item and Function	Potential Failure Mode	Potential Effect(s) of Failure	Severity of Failure	Potential Cause(s) of Failure	Occurrence (Probability of Failure)	Current Controls	Detection (Ability to Detect)	RPN

FMEA expands to calculate the effects of failures in terms of a risk priority number (RPN), which is also a function of risk factors such as severity, occurrence, and detection.

$$PRN = Severity \times Occurrence \times Detection$$

Item and Function	Potential Failure Mode	Potential Effect(s) of Failure	Severity of Failure	Potential Cause(s) of Failure	Occurrence (Probability of Failure)	Current Controls	Detection (Ability to Detect)	RPN	Recommended Actions	Responsibility and Target Completion Date

Stage 3. Development and Identification of Contingency Plans

Examples of an FMEA and a risk prioritization number are given in the Chapter 9 case study similar to the following one. They show before and after actions taken to reduce the defects.

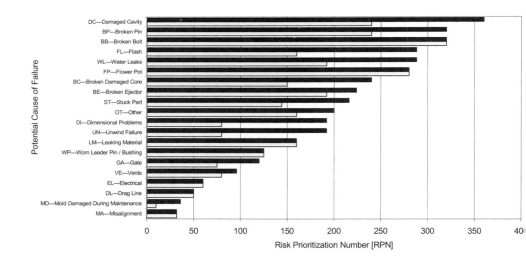

- Rating of RPN

The RPN rating is given in Table 8.6.

Furthermore, an RPN rating in terms of process capability is given in Table 8.7.

Table 8.6

Rating of Risk Priority Number

Rating	Description	Definition
10	Dangerously High	Failure could injure the customer or an employee
9	Extremely High	Failure would create noncompliance with federal regulations
8	Very High	Failure renders the unit inoperable or unfit for use
7	High	Failure causes a high degree of customer dissatisfaction
6	Moderate	Failure results in a subsystem or partial malfunction of the product
5	Low	Failure creates enough of a performance loss to cause the customer to complain
4	Very Low	Failure can be overcome with modifications to the customer's process or product, but there is a minor performance loss
3	Minor	Failure would create a minor nuisance to the customer, but the customer can overcome it in the process or product without performance loss
2	Very Minor	Failure may not be readily apparent to the customer but would have minor effects on the customer's process or product
1	None	Failure would not be noticeable to the customer and would not affect the customer's process or product

And an additional analysis regarding inspection rating of the risk prioritization number is given in Table 8.8.

Improving the RPN

To minimize severity, one must change the design. Likewise, to minimize occurrence, design or process change is required. In the same way, to minimize detection, one should move to less operator-dependent verifications/inspections.

Summary of FMEA Steps

1. Find the product, process, or system components.
2. Brainstorm potential failure modes.
3. List potential effects of each failure mode.
4. Specify a severity rating for each effect.

Table 8.7

Rating in Process Capability

Rating	Description	Definition
10	Very high: Failure is almost inevitable	More than one occurrence per day or a probability of more than three occurrences in 10 events (Cpk < 0.33)
9		One occurrence every three to four days or a probability of occurrence in 10 events (Cpk < 0.67)
8	High: Repeated failures	One occurrence per week or a probability of five occurrences in 100 events (Cpk < 0.83)
7		One occurrence every month or one occurrence in 100 events (Cpk < 0.83)
6	Moderate: Occasional failure	One occurrence every three months or three occurrences in 1,000 events (Cpk = 1.00)
5		One occurrence every six months to one year or three occurrences in 10,000 events (Cpk = 1.17)
4		One occurrence per year or six occurrences in 100,000 events (Cpk = 1.33)
3	Low: Relatively few failures	One occurrence every one to three years or six occurrences in 10 million events (Cpk = 1.67)
2		One occurrence every three to five years or two occurrences in 1 billion events (Cpk = 2.00)
1	Remote: Failure is unlikely	One occurrence in greater than five years or less than two occurrences in 1 billion events (Cpk > 2.00)

5. Specify an occurrence rating for each effect.
6. Specify a detection rating for each failure mode or effect.
7. Perform the RPN calculation.
8. Prioritize the failure modes for action.
9. Take action to reduce the high-risk failure modes.
10. Calculate the resulting RPN.
    - **The Laws of Lean *Six Sigma***
    See Chapter 6.

Once the process data has been analyzed, the results will indicate the dominant factors that cause process defects. Then the process will be simulated or optimized through optimization theory and techniques, such as statistical methods,

Table 8.8
RPN Inspection Rating

Rating	Description	Definition
10	Absolute Uncertainty	The product is not inspected; or the defect caused by failure is not detectable
9	Very Remote	Product is sampled, inspected, and released on acceptable quality level (AQL) sampling plan
8	Remote	Product is accepted based on no defects in a sample
7	Very Low	Product is 100% manually inspected in the process
6	Low	Product is 100% manually inspected using go/no-go or mistake-proofing gauges
5	Moderate	Some statistical process control (SPC) is used in the process, and the product sets a final inspection offline
4	Moderately High	SPC is used, and there is immediate reaction to out-of-control conditions
3	High	An effective SPC program is in place with process capabilities (Cpk) greater than 1.33
2	Very Low	All product is 100% automatically inspected
1	Almost Certain	The defect is obvious, or there is 100% automatic inspection with regular calibration and preventive maintenance of the inspection equipment

analysis of variance (ANOVA), design of experiment, and other methods (refer to process improvement methods).

Absolute Quality Production (AQP)/Process Improvement

At present, almost all the production in any organization depends on the operating equipment. Indeed, even the quality of many processes relies on machines—either manufacturing instruments or computers. Therefore, defect free is impossible without complete prevention maintenance.

8.6 EVALUATION OF NEW SIGMA

As in Section 8.3, compute the following list of parameters:

- Determine defect-free probability (DFP) or defect per million opportunities (DPMO) for the new data.
- DPMO = [(Total defects)/(Total opportunities)] $* 10^6$
- Determine the sigma shift, μ, σ, C_p, C_{pk}, Gage R&R, and so on.

- Calculate the process sigma rating from the defect-free probability.
- Determine the z-value or sigma capability.

Review Microsoft Excel function "= normsdist (z), = normsinv (probability)."

8.7 PROCESS CONTROL

By definition process control means to ensure that the process will produce the same products all the time. In addition, articles having the same physical properties and functionality will meet customer requirements. This step requires a process and design control plan and team development, as well as SPC and advanced process control techniques.

8.7.1 PHASE IV: PROCESS CONTROL AND MAINTAIN

As described in Chapter 3, the process output (Ys) is a function of process input (Xs)—that is, Output = f (input) or $Y = f(X)$.

where Y is called dependent, output, effect, symptom, or monitor, and X is called independent, input, cause, problem, or control.

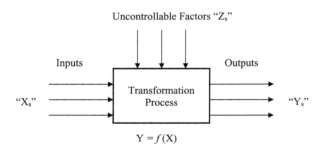

Questions during process control may arise, such as the following:

Question 1	What are the most effective Xs in the process?
Question 2	What are good nominal targets of X values?
Question 3	What is the expected tolerance/variation of X?
Question 4	Is X value measurable?
Question 5	What actions should be taken if X values deviate from the desired set point?
Question 6	Will the team continue to monitor the primary output (Ys) matrix? If yes, then
Question 6.1	How often?

Question 6.2   What actions should be taken if Y values deviate from expected values?

Question 6.3   Must any uncontrollable parameters (Z factors) be measured, and, if yes, what methods will be used?

Positrol

Positrol comes from **Posi**tive (Posi) and Con**trol** (trol)—the important input variables controlled in day-to-day production. The control process is defined as follows:

a. *Who* will monitor, measure, and record the process?
b. *How* will instruments or tools be used in measuring the process?
c. *Where* will the process take place?
d. *When* will measurement take place?

A positrol control plan includes the following:

- Specific aspects of the process that must be controlled.
- Who is responsible to maintain control, measure, check, replace, and apply mistake proof?
- Where is the location of the workstation?
- When are the time intervals, what is the number of pieces, and what is the frequency of measurement?
- How will process control be measured and with what techniques, gauges, procedures, and devices?

Once the solutions are selected, a control plan must be developed to ensure that improvements will be sustained. Process improvement will not be sustained if the control plan is not followed. A process control plan is a one-page summary of the control plan.

In this phase the process control, process maintenance, implementation of the monitoring system, and documentation will be reviewed.

Process Control Objectives

a. Implement the process control planning system.
   i. This will ensure that customer requirements are in control, and process knowledge can be passed on to new process team members through time.
   ii. Illustrate process flow.
   iii. Document key process metrics, including their measurement frequency, specification, or target.
   iv. Document responsibilities of teams.
   v. Identify actions in the event of process upsets.
b. Implement control charts for key variables.

   c. Implement lean tools and lean enterprise.
   d. Apply mistake-proofing techniques.
      i. Mistake proofing is an effective technique for eliminating and preventing defects.
      ii. It is a specific tool for specific applications that can be applied to products, process, and tool designs.
      iii. It is a simple cost-effective method to minimize defect rates and increase quality.
   e. Interpret control charts and evaluate results.
   f. Establish a documentation plan. All the project's actions and results should be documented with all the supporting evidence for successful operation.
   g. Monitor performance metrics.
      i. Create control/monitoring plan.
      ii. Determine input/output ongoing measurement system.
      iii. Find how input, process, and output variables will be checked.
      iv. Apply control charts.
      v. Consider if any training is required for control charts' interpretation.
      vi. What is the most recent process yield?
      vii. Does the process performance meet the customer's requirement?
   h. Risk prioritization number before and after.
      • Failure mode effect analysis before and after.

Process Control Tools and Techniques

• Process control plan

Establish and evaluate a process control system (statistical process control). Then monitor process response and maintain the control of the process inputs (variables) under SPC guidelines. Some examples of control plans are given in Figures 8.12–8.14.

• Control/precontrol chart (ongoing)

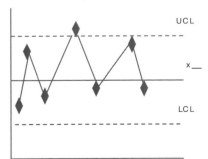 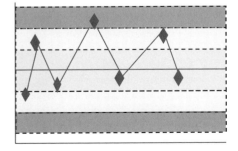

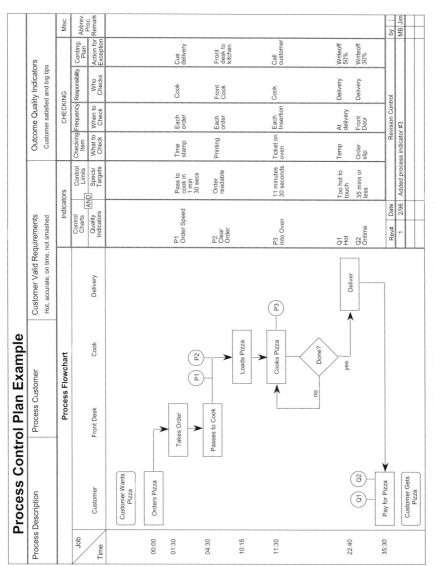

Figure 8.12 Process control plan for pizza delivery.

163

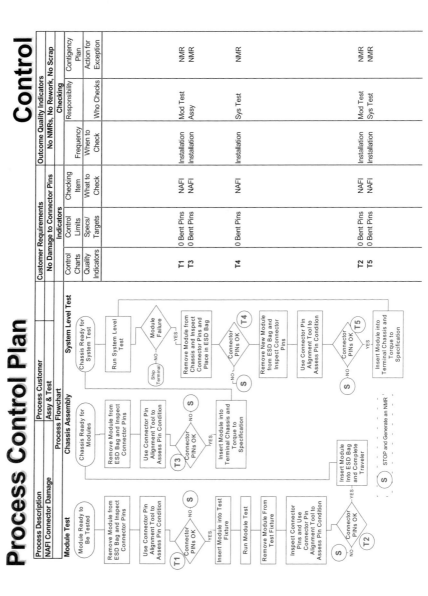

Figure 8.13 Process control plan for NAFI connector damage.

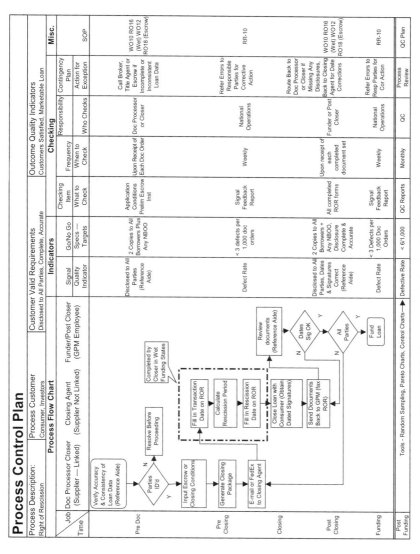

Figure 8.14 Process control plan for right of rescission (loan).

Originally, this control chart was introduced by Dr. Walter A. Shewhart in the 1920s. It is often referred to as the "Shewhart chart." The objectives of control charts are to enable the front-line worker to distinguish between random variation and variation due to an assignable cause and to monitor process performance over time for stability. It also helps to identify opportunities and to understand and control variations.

The common types of control charts depend on the circumstance and type of data available to determine and construct the following charts:

- $\bar{X}$—average
- R—range
- $\bar{X}$ and R—average and ranges
- $\bar{X}$ and S—average and standard deviation

Mathematically, models for the $\bar{X}$ center line and control chart limits using R are defined as follows:

$$UCL = \bar{\bar{X}} + A_2 \bar{R}$$
$$\text{Center line (CL)} = \bar{\bar{X}}$$
$$LCL = \bar{\bar{X}} - A_2 \bar{R}$$

Comments: Value of A_2 is given in Table 8.9.

Where, $\bar{\bar{X}} = \dfrac{\sum \bar{x}_i}{m}$ and $i = 1, 2, \ldots, m$ is X double bar (Equation 8.32)

process mean (an estimator of population mean), and $\bar{R}$ is called R-bar (Equation 8.33).

Sample	X-bar	R
1	29.1247	3.9354
2	30.9047	7.6413
3	28.5579	5.6472
4	28.0154	4.6327
5	32.8012	7.8571
6	31.3201	4.8216
7	30.4921	5.7591
8	29.3421	5.0367
9	30.7821	2.4435
10	29.6043	7.6483
11	31.5065	6.5534
12	31.1112	5.0572
13	28.7819	6.7418
14	30.5546	6.2371
15	31.7024	4.7026

EXAMPLE 8.8.   An engineer was assigned to evaluate the parts weight of an experiment with 15 runs, each run with 4 samples. The resultant $\bar{X}$ (mean of subgroup) and R (range of subgroup) are given here:

$$\bar{\bar{X}} = \frac{\bar{X}_1 + \bar{X}_2 + \cdots + \bar{X}_{m-1} + \bar{X}_m}{m} \tag{8.32}$$

$$\bar{\bar{X}} = \frac{29.1247 + 30.9047 + \cdots + 30.5546 + 31.7024}{15} = 30.30675$$

$$\bar{R} = \frac{R_1 + R_2 + R_3 \cdots + R_{m-2} + R_{m-1} + R_m}{m} \tag{8.33}$$

$$\bar{R} = \frac{3.9354 + 7.6413 + \cdots + 6.2371 + 4.7026}{15} = 5.64767$$

Using Table 8.9 at n = 4, the control limit factor is A_2 = 0.729.

$$\text{UCL} = \bar{\bar{X}} + A_2\bar{R} = 30.30675 + (0.729)(5.64767) = 34.4239$$

where, $A_2 = \dfrac{3}{d_2\sqrt{n}}$

Table 8.9
The Control Factors for Constructing Control Charts

n	A_2	D_3	D_4	d_2	d_3
2	1.880	0.000	3.267	1.128	0.853
3	1.023	0.000	2.574	1.693	0.888
4	0.729	0.000	2.282	2.059	0.880
5	0.577	0.000	2.115	2.326	0.864
6	0.483	0.000	2.004	2.534	0.868
7	0.419	0.076	1.924	2.704	0.833
8	0.373	0.136	1.864	2.847	0.820
9	0.337	0.184	1.816	2.970	0.808
10	0.308	0.223	1.777	3.078	0.797
11	0.285	0.256	1.744	3.173	0.787
12	0.266	0.284	1.716	3.258	0.777
13	0.249	0.308	1.692	3.336	0.769
14	0.235	0.329	1.671	3.407	0.762
15	0.223	0.348	1.652	3.472	0.754
16	0.212	0.364	1.636	3.532	0.749
17	0.203	0.379	1.621	3.588	0.743
18	0.194	0.392	1.608	3.640	0.738
19	0.187	0.404	1.596	3.689	0.733
20	0.180	0.414	1.586	3.735	0.729

Center line $(CL) = \bar{\bar{X}} = 30.3068$

$LCL = \bar{\bar{X}} - A_2 R = 30.30675 - (0.729)(5.64767) = 26.1896$

Likewise, models for the R control chart limits are as follows:

$$UCL = D_4 \bar{R}$$
$$CL = \bar{R}$$
$$LCL = D_3 \bar{R}$$

Values of D_3 and D_4 are defined as given in Equation 8.34:

$$D_3 = 1 - 3\frac{d_3}{d_2} \text{ and } D_4 = 1 + 3\frac{d_3}{d_2} \qquad (8.34)$$

where d_2 is the factor for estimating sample standard deviation.

$$s = \frac{\bar{R}}{d_2}$$

values of d_3 and d_2 are provided in Table 8.9. Since sample range values are not negative, D_3 values are zero whenever Equation 8.34 is negative—that is, when n = 2, 3, 4, 5, and 6. In addition, values of D_3 and D_4 also are given in Table 8.9.

For example, 8.6 at n = 4 from Table 8.9, $D_4 = 2.282$ and $D_3 = 0.0$, so the outcome will be

$$UCL = D_4 \bar{R} = (2.282)(5.64767) = 12.888$$
$$CL = \bar{R} = 5.64767$$
$$LCL = D_3 \bar{R} = 0$$

Similarly, control models for the $\bar{X}$ center line and control chart limits using S are defined as

$$UCL = \bar{\bar{X}} + A_3 \bar{S}$$
$$CL = \bar{\bar{X}}$$
$$LCL = \bar{\bar{X}} - A_3 \bar{S}$$

Summary of Procedure for Creating $\bar{X}$ and R Charts

1. Collect m samples of data m sets, each size of n (i.e., m = 15 data sets, each set equal to a sample size of 5).
2. Calculate the mean of each subgroup (i.e., $\bar{X}_1, \bar{X}_2, \ldots \bar{X}_{m-2}, \bar{X}_{m-1}, \bar{X}_m$).
3. Calculate the range of each subgroup (i.e., $R_1, R_2, \ldots R_{m-2}, R_{m-1}, R_m$).
4. Determine the overall mean for step 2, $\bar{\bar{X}}$, where $\bar{\bar{X}}$ is the mean of m sets of $\bar{X}$.

5. Determine the overall mean for step 3, $\bar{R}$, where $\bar{R}$ is the mean of m sets of R.

6. Calculate $s = \dfrac{\bar{R}}{d_2}$ to estimate s, where values of d_2 are listed in Table 8.9.

7. Calculate the three sigma control limits for the $\bar{X}$ control chart:

$$\text{UCL} = \bar{\bar{X}} + 3\frac{s}{\sqrt{n}} = \bar{\bar{X}} + 3\frac{\left(\dfrac{\bar{R}}{d_2}\right)}{\sqrt{n}} = \bar{\bar{X}} + \frac{3}{d_2\sqrt{n}}\bar{R} = \bar{\bar{X}} + A_2\bar{R}$$

Center line $(\text{CL}) = \bar{\bar{X}}$

$$\text{UCL} = \bar{\bar{X}} - 3\frac{s}{\sqrt{n}} = \bar{\bar{X}} - 3\frac{\left(\dfrac{\bar{R}}{d_2}\right)}{\sqrt{n}} = \bar{\bar{X}} - \frac{3}{d_2\sqrt{n}}\bar{R} = \bar{\bar{X}} - A_2\bar{R}$$

8. Calculate the three sigma control limits for the $\bar{X}$ control chart:

$$\text{UCL} = D_4\bar{R}\ (\text{refer to Table 8.9})$$
$$\text{CL} = \bar{R}$$
$$\text{LCL} = D_3\bar{R}\ (\text{refer to Table 8.9})$$

9. Plot the control chart using the $\bar{X}$ and R points for each subgroup on the same vertical line for sample numbers.

Control Limits versus Specification Limits

Often, there is confusion between control limits and specification limits. In fact, not only are they not related, but they are also different. In other words, the control limits are dependent on the experimental data outcome, but the specification limits are based on the customer desire (or product functionality) limits. For a product to be cheaper and more profitable, specification limits need to be outside the control limits. The comparison of control limits versus specification limits is given in Table 8.10.

- Poka-Yoke (Error-Proofing)
  Poka-Yoke also called mistake proofing. It is a process designs that are robust to mistakest (error free). The mistake-proofing steps are identifying the problem, prioritizing the problem, finding the root cause, creating solutions, and measuring the results.
- Apply an ongoing Pareto chart.
  This identifies the 20% of sources that cause 80% of the problems and focuses the efforts on the "vital few."

Table 8.10

Control Limits versus Specification Limits

Control limits	Specification limits
Process data	Customer specifications
UCL, CL, LCL must be calculated using data collected from the process	USL, LSL are defined in the part blueprints based on the customer requirements
It is on the control charts	It appears on the histogram
Data is used to determine if the process is in statistical control (e.g., process is consistent)	It is used to determine if the process is outside the specification limits (e.g., product will function within the specified limits)
If the points of graphs are outside the control limits, then the process is not in statistical control	If the points of graphs are outside the specification limits but within the control limits, then the process is in statistical control, but it is not capable of meeting customer requirements

- Continue ongoing process capability, FMEA before and after improvement, as well as RPN.
- Make no mistake, and increase your accuracy.
- Establish and evaluate a process control system.
- Monitor process response and maintain the control of the process inputs (variables) under SPC guidelines.
- Continue applying Lean manufacturing methods.
- Maintain control and monitoring performance of metrics.

The discussion in this chapter is an approach to present *Six Sigma* effectiveness to the design of various processes in the organization. To ensure that the customer will receive products that will function well, *Six Sigma* must be applied continuously and effectively.

The interaction of *Six Sigma* departments in the company is shown in Figure 8.15. Benchmarking should be included in all steps of the process. As mentioned in Chapter 1, Lean *Six Sigma* is a five-year program. The fifth year is the *Six Sigma*. Each year, as the sigma level increases, the process gets more challenging and more analytical.

A flowchart of a Lean *Six Sigma* problem-solving strategy is shown in Figure 8.16.

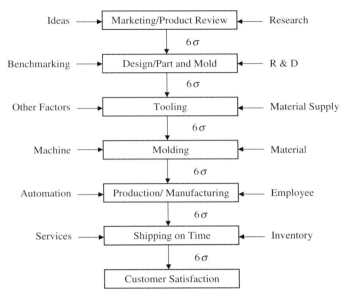

Figure 8.15   Departmental interaction of company on the fourth year Lean *Six Sigma* program.

Overview Summary of *Six Sigma* in the Entire Organization

The following lists are the overall organizational focus to achieve *Six Sigma* concepts and goals.

- **Organizational Focus**

  The staff of an organization should understand the philosophy of *Six Sigma*, problem-solving technique (roadmap), as well as benchmarking and long-term business goals to achieve the highest quality. Furthermore, as a team they need to share common goals and continue to improve as well as reduce defects in all business operations by involving everyone in improvement efforts. Some of the important continuous activities on the projects and responsibilities are as follows:

  1. On or ahead of time
  2. On or ahead of schedule
  3. On or under budget
  4. On or under cycle time (production yield, inventory)
  5. On or ahead of delivery
  6. Quality (reduced defects)
  7. Be able to meet all the targets as assigned

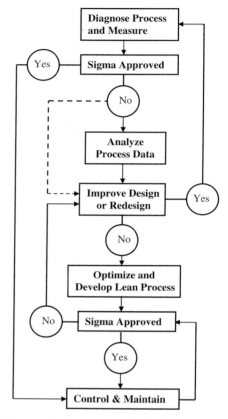

Figure 8.16   Flowchart of the *Six Sigma* concept.

- **Product/Tooling Design**
  The elements of product and tool design to achieve a robust finished product list are as follows:
  - Understanding of *Six Sigma* robust design
  - Creative *Six Sigma* vision
    - Design concept review
    - Statistical tolerance study
    - Manufacturability, processability, functionality, and cost affectivity
    - Raw material quality and specification
    - Application of optimizing tools and utilities
    - Ergonomic/failure study

Table 8.11

Summary of Black Belt Lean *Six Sigma* Statistical Tools

- Lean
  - Pull, value stream, lead time (speed), and flow
- Analysis
  - Analysis of covariance
  - Analysis of variance
  - Worse-case analysis
  - Failure mode analysis
  - Regression analysis
    - Multiple linear regression
    - Nonlinear regresssion
- Descriptive graphs
  - Histograms, frequency polygons, cumulative frequency, bar, pie, and Pareto charts
  - Run/line
  - X-bar (average), R (range), S (standard deviation), np (number nonconforming), p (proportion nonconforming), U (defect per unit), process control (planning and implementing), and SPC (statistical process control) charts
- Cause-and-effect matrices
- Correlation
  - Correlation and simple linear regression
  - Correlation and multiple linear regression
- Descriptive measures
  - Variation, position, shape, and box plots
- Design of experimental techniques and analysis
  - 2^2 Full factorial design and analysis
  - 2^3 Full factorial design and analysis
  - 2^k Full factorial design and analysis
  - 2^{4-1} Fractional design and analysis
  - 2^{5-1} Fractional design and analysis
  - 2^{6-2} Fractional design and analysis
  - 2^{k-p} Fractional design and analysis
  - Response surface design and analysis (statistical methodology in optimizing the process)
  - Central composite design (see case study in Chapter 10)
  - Robust design
- Distributions
  - Chi-square analysis, binomial, F-distribution, frequency distribution, exponential distribution, normal (continuous), Poisson (discrete), t-distribution, and z-distribution (standard normal distribution)
- Hypothesis testing
- Point estimate (estimation of population using analysis of samples drawn from the population)
- Process capability
- Statistical inferences and sampling
  - Confidence intervals
  - Random sampling
- Statistical process control
- Tests: F-test, t-test, z-test, Chi-square (χ^2), goodness-of-fit tests, and Chi-square tests of independence
- Other tools, such as mean, variance, histogram, continuous (normal) distribution curve, discrete (Poisson) distribution, the standard deviation (sigma), point estimate (estimation of population using analysis of samples drawn from the population), and hypothesis testing. In addition, statistical samplings, analysis of variance, design of experiment (single, two, and multifactorial design, fractional factorial design), worst-case analysis, failure mode analysis, multiple linear, and nonlinear regression

- **Production Process**
  - Be able to translate process information into statistical information and model.
- **Finished Products**
  - Delivery (on-time)/quality/customer satisfaction

Chapter 9

Six Sigma Green and Black Belt Level Case Studies

9.1 CASE STUDY: METHODOLOGY FOR MACHINE DOWNTIME REDUCTION—A GREEN BELT METHODOLOGY

9.1.1 PHASE 0: PROBLEM STATEMENT

The automation system of a manufacturing company X during the fall quarter had over 45% downtime, resulting in an additional labor cost of $140,000–$145,000. If this trend continues within a year, the labor cost will exceed $500,000. A team of engineers and production staff were assigned to study the case. The objective is to reduce total downtime by at least half to a maximum of 22.5% and not to exceed that thereafter. This would allow customer demand to be met without the requirement of overtime and about 100,000 parts per week, which makes about 40% of X product.

Approach

First, the Lean *Six Sigma* team began with a process map SIPOC and flow-chart, as shown in Figures 9.1 and 9.2, to clearly understand the process itself.

9.1.2 PHASE 1: DATA COLLECTION AND MEASUREMENT

All process outputs relevant to downtime are attributing in nature and recorded in a networked database. Some of the issues that cause machine downtime are as follows:

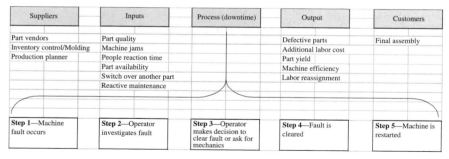

Figure 9.1   Process mapping.

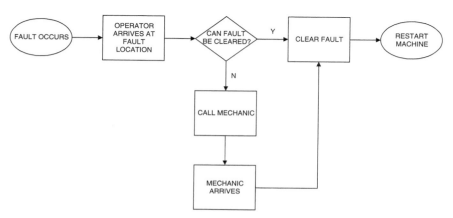

Figure 9.2   Process flowchart.

- **Fault description and frequency**
  Machine data detailing all faults.
- **Lack of parts**
  Downtime accrued when the automation system does not run due to the lack of availability of a particular part in the assembly.
- **Jam parts**
  Downtime accrued when the automation system stops for a machine station fault until the time the machine is restarted.
- **Reactive maintenance (RM)**
  Downtime accrued during the time maintenance assists in the clearing of a fault due to machine error.
- **Preventive maintenance (PM)**
  Time accrued while maintenance performs required PM on the automation system.

- **Quality issues**

  Downtime accrued when a part defect causes the automation system to stop.
- **Switch over to different part (changeover)**

  Downtime accrued during a product or part change.
- **Mixed parts**

  Downtime accrued when multiple parts are mixed and entered into the automation system.

Furthermore, to establish a performance baseline for measurement and improvements, the following process capability calculations data validation chart (Figure 9.3) and Pareto chart of process variation (Figure 9.4) were developed.

$$\text{Maximum capacity of automation system} = 132.6E3 \text{ units/week}$$
$$\text{Desired output} = 1.0E5 \text{ units/week}$$
$$1.0E5/132.6E3 = 77\% \text{ required yield}$$
$$(85 \text{ total hrs. available per week}) \times (1 - 0.77) =$$
$$19.55 \text{ hours allowable downtime/week}$$

The downtime tolerance is 1.95 hours per shift (USL) or 19.55 hours per week. Yield versus desired limits are as followed:

$$\text{Defect per opportunity (DPO)} = D/(N \times O)$$
$$\text{Yield } \% = (1 - \text{DPO}) \times 100$$
$$D = 289.4 \text{ hours, } N = 625 \text{ hours, } O = 1 \text{ hour}$$
$$\text{DPO} = 289.4/(625 \times 1) = 4.63E{-}1$$
$$\text{Yield } \% = (1 - 4.63E1) \times 100 = 53.7\%$$

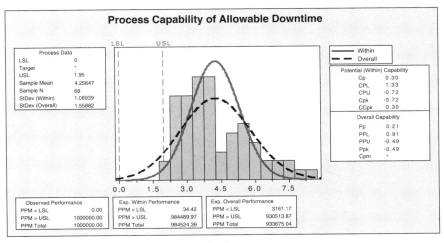

Note: Data samples taken from two months data.

Figure 9.3 Data validation.

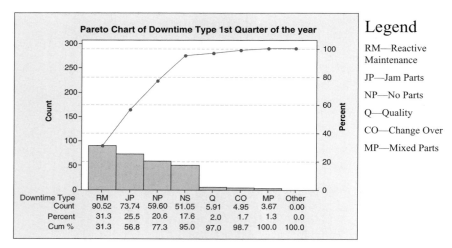

Figure 9.4 Pareto chart of process variation.

Based on the existing Sigma level:

$$DPMO = 4.63E5$$

Process sigma capability was calculated using the concept from Chapter 2:

$$\text{NORMSINV}\left(1 - \frac{defects}{1.0E6}\right) + 1.5 = 1.593 \cong 1.6 \text{ Sigma}$$

or see sigma conversion table in the Appendix.

Thus, based on the information discussed here, the project objectives are established to achieve the following goals:

1. Increase efficiency of automation equipment.
2. Increase productivity and utilization.
3. Increase quality → decrease scrap.
4. Decrease operating expenses.
5. Eliminate waste.
6. Improve Takt time for all equipment.
7. Establish headcount requirements (all shifts).
8. Implement in-process/finished good inspectors.
9. Establish performance measures.
10. Create project schedule.

9.1.3 PHASE 2: ANALYSIS OF MEASUREMENT

Data obtained from phase 1 were analyzed by the team to find the improvement opportunities. Many tools were used in this process, but some of the highlights are illustrated in Figures 9.5–9.8.

The Pareto chart also was applied in identifying the top five faults of this project.

9.1.4 PHASE 3: IMPROVE AND VERIFY ANALYZED DATA

A benchmarking was established as follows:

- Automation issues (major faults)
- In-process inspection
- Automation measurements
- Scrap costs evaluation
- Finished goods completion
- Historical data review
- Labor routing analysis

In addition to benchmarking, further process improvements introduced were as follows:

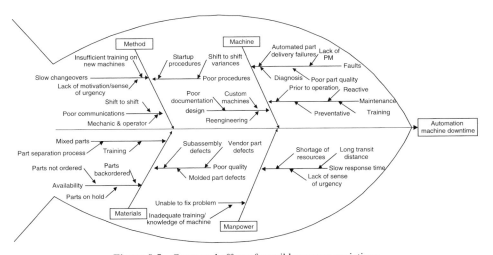

Figure 9.5   Cause and effect of possible process variations.

THEME	Automation Feeder Faults	Automation Machine fault	Changeover	Reaction Time	Part Quality
IDEAS	Par A Jams (9)	Part N Faults (15)	Part B wire spool (10)	Operator Response (4)	Molded Part Quality (4)
	Part B Jams (9)	X Welder Faults (9)	Product Change Over (3)	Training (4)	Bad Sub-Assembly. (3)
	Part A Presents (8)	Software Issues (7)	No Spare Parts Avail. (2)	Maintenance Response (3)	
	part B Presents (6)	Part D Faults (6)		Machin Y Re-start (1)	
	Part M (6)	Part M Faults (5)		Low Labor Availability (1)	
	Part C Presents (5)	Part C Fault (3)		Bad Machine # 2 Monitor Location (1)	
	Part D Jams (5)	Part E Faults (2)			
	Part E Jams (5)	Part K Cover (1)			
	Machine# 2 Transfer (4)				
	Part F Presents (3)				
	Part G Presents (2)				
	part H jams (2)				
	Mixed Parts (2)				
	Part K Presents (2)				

Note: Numbers in paratheses represent number of times variable was mentioned.

Figure 9.6   Affinity diagram/brainstorming.

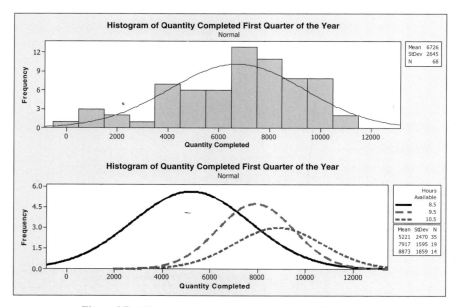

Figure 9.7   Histogram of completed products for the first quarter.

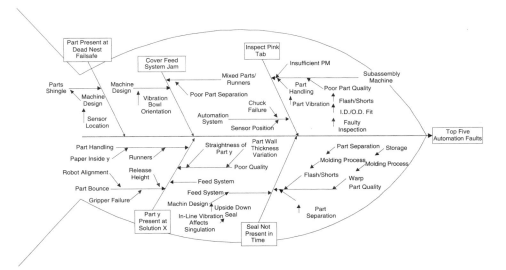

Figure 9.8   Cause-and-effect diagram of top five faults.

- Modification and upgrading of automation equipment
- Necessary changes to company X product, such as
  - Process and marketing specifications
  - Design changes
- Apply Lean manufacturing principles
  - Kanban
  - Status charts
  - *Six Sigma* tools (reducing variation)

The Phase 3 investigation using analysis of variance (Figure 9.9) indicated that that data were in normal distribution. Further, downtime is reduced as the number of operators increased from one to four. But the downtime began to increase by implementing an additional operator, as shown in the box plot in Figure 9.10.

Other tools include: FMEA, RPN before and after improvement, and time series plot (occurrence versus time). Here are some other applications in addition to the tools discussed:

- Increased orientation specification on machine
- Added Poka-Yoke gauge to molding inspection
- Evaluated part handling process from subassembly machine
- Incorporated controls for scrap
- Initiated control chart for fault occurrence

Hypothesis Test:
Ho: Shift 1 not equal Shift 2   Significant Variable = 0.05
H1: Shift 1 not equal Shift 2   F-Test: Variances are equal

Two-Sample T-Test:
Reject HO   Shift 1 and Shift 2 are statistically different

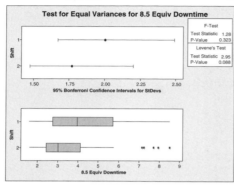
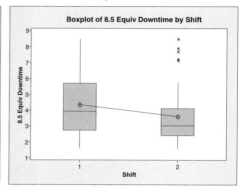

Figure 9.9   Comparison of statistical analysis (box plots) for shifts 1 and 2.

One-Way ANOVA: (8.5 Equivalent Downtime Hours versus Number of Operators)

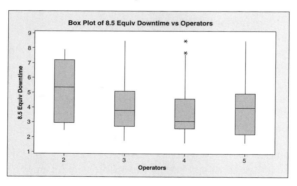

Figure 9.10   Box plot analysis of operators in active and downtime.

9.1.5 PHASE 4: CONTROL AND MAINTAIN

These Lean *Six Sigma* activities are adapted to continuously improve, control, and maintain the established process:

- Improve data collection methods
- Work instructions

- Control plans
- 5S methodology
- Changeover time
- Total productive maintenance (TPM)
- Prolong productivity performance (PPP)
- Spare parts inventory
- Mistake proofing (Poka-Yoke)
- Potential Failure Mode Effects Analysis (PFMEA)
- Training new employees (rotation system)
- Process flow diagram
- Reusable WIP bins

These are additional monitoring plans in place:

- Review statistical data of machine faults/downtime data at weekly team meetings
- Create real-time monitoring of machine fault occurrences using control charts (e.g., $\bar{X}$ and R).
- Create/track network-based action items list to be reviewed at weekly team meetings
- Monitoring chart for individual machine downtime (Figure 9.11)

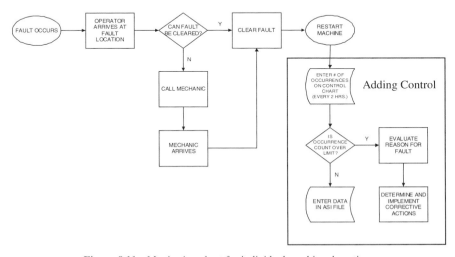

Figure 9.11   Monitoring chart for individual machine downtime.

9.2 CASE STUDY: METHODOLOGY FOR DEFECT REDUCTION IN INJECTION MOLDING TOOLS—A BLACK BELT METHODOLOGY

9.2.1 PHASE 0: DEFINITION AND STATEMENT OF ISSUES

A large manufacturing company with an injection molding facility in the United States had significant tool (see Figure 9.12) repair expenses. The report showed that total tool repair cost exceeded over U.S. $2 million in one year. Obviously, there were no data for analysis. Thus, senior managers decided to clean up and establish a systematic system to reduce the cost by 20% in the first year and continue to reduce thereafter.

Goal: To reduce cost by at least 20% or more and continue to reduce cost and maintain it thereafter.

9.2.2 PHASE 1: DATA COLLECTION AND MEASUREMENT

Process of Data Collection and Analysis

The project started with overall process scope SIPOC (**S**upplier, **I**nput, **P**rocess, **O**utput, and **C**ustomer), as shown in Figure 9.13. Project was defined on cost reduction, and data were collected. Since cost reduction had no specific limit of target, the customer's desire was to reduce the cost as much as possible. Therefore, in this case LSL was set to zero (optimum point/zero defects). Data was collected (baseline data) for the existing process, and process capability was estimated using the computer software Minitab. This is shown in Figure 9.14.

The four steps of Lean *Six Sigma* methodologies were applied. Lean *Six Sigma* tools were implemented where they were required.

The "house of quality" analysis established in Table 9.1 shows the QFD matrix with rating scores of *how* to achieve technical solutions (see Chapter 8). However, the results shown in Table 9.2 are the outcome of the *technical solution ratings* multiplied by the *customer importance rating factors* in Table 9.1 to come up with a final rating score. Hence, absolute scores are the sum of the individual measurement in each column. The following example illustrates the calculation of the concept in Tables 9.1 and 9.2.

$$\text{Relationship matrix score} = (\text{Customer importance rating})$$
$$\times (\text{Technical solution rating})$$

That is, in the case of "less down time" and "work experience," using Table 9.1, the customer importance rating is equal to five, and the technical solution rating

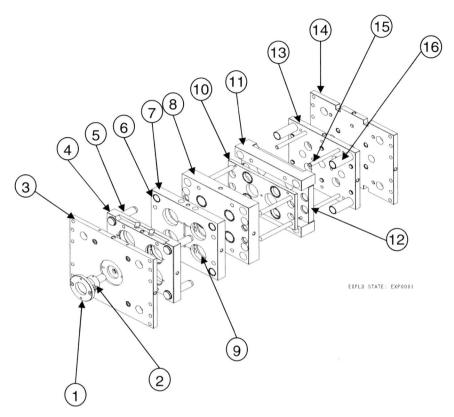

1. Locating Ring
2. Sprue Bushing
3. Front Clamping Plate
4. Front Cavity Plate ("A" Plate)
5. Leader Pins
6. Leader Pin Bushings
7. Rear Cavity Plate ("B" Plate)
8. Support Plate

9. Cavity Bore
10. Ejector Retainer Plate
11. Spacer Block
12. Ejector Plate
13. B-Core Plate
14. Rear Clamping Plate
15. Screw to Hold B-Half Together
16. Guide Sleeve

Figure 9.12 Exploded view of mold.

Table 9.1

House of Quality: Partial Relationship (Quality Function Deployment)

Customer Requirements	Whats	Rating Factor	Equipment	Training			Documentation		Staffing		Organization		Prevention		Communication		
			Machinery	Formal Classes	Work Experience	Company Classes	Accurate Prints	Appropriate Paperwork	Quantity	Expertise	Workflow	Inventory Control	Mold Handling	Inspection	Project Status	Problem Identification	Possible Solution
Fast Turnaround	Less Downtime	5	9	3	9	3	9	9	9	9	9	9	3	3		9	3
Reliable Tooling	Robust Design	5		3	9	3	1			3				1		3	3
	Steel Type	2		3	3	3	1			3				1		3	
	Wear Factors	2		3	3	3	1			3				1		3	1
	Proper Maintenance	5	3	3	9	3	1		3	3	3	9	1	1		9	3
Accuracy	Dimensional Accuracy	5	9		3		9	9		3				9		1	
	No Flash	5	3	3	3					3				3		1	3
Mold Improvement	Cooling	3	3	3	9	3				3				3		3	3
	Venting	3	3	3	9	3				3				3		3	3
	Runner System	3	3	3	9	3				3				3		3	3

Category	Requirement															
	Gating	3	3	3	9	3				3				3		3	3
	Ejection	3	3	3	9	3				3				3		3	3
	Mold Component Life	5	3	3	9	3				3			3	3		9	3
Communication	Repair Estimate	4			3		3	9	1	3				3	3	3	3
	Repair Completion	5					3	3	1	1				3	3	1	
	Work Performed	3							1	1							3
Effective Tooling	Hot Runner	3	3	3	3				1	1							
	Cold Runner	3	3	3					1	1							
	Optimum Cycle Time	5		3	9				3	3					3		
	Ease of Setup	4															
	No Water Leak	5		3	3				3	3							
Technical Importance	Absolute		48	39	105	45	28	30	15	58	12	18	7	43	6	60	34

Table 9.2

House of Quality: Scored (Quality Function Deployment)

Customer Requirements	Whats (Solutions / Hows)	Rating Factor	Equipment	Training			Documentation		Staffing		Organization		Prevention		Communication		
			Machinery	Formal Classes	Work Experience	Company Classes	Accurate Prints	Appropriate Paperwork	Quantity	Expertise	Workflow	Inventory Control	Mold Handling	Inspection	Project Status	Problem Identification	Possible Solution
Fast Turnaround	Less Downtime	5	45	15	45	15	45	45	45	45	45	45	15	15	0	45	15
			0														
Reliable Tooling	Robust Design	5	0	15	45	15	5			15				5		15	15
	Steel Type	2	0	6	6	6	2			6				2		6	2
	Wear Factors	2	0	6	6	6	2			6				2		6	15
	Proper Maintenance	5	15	15	45	15	5		15	15	15	45	5	5		45	15
			0														
Accuracy	Dimensional Accuracy	5	45	3	15		45	45		15				45		5	
	No Flash	5	15	9	15					15				15		5	
			0														
Mold Improvement	Cooling	3	9	3	9	27	9			9				9		9	9
	Venting	3	9	9	27	9				9				9		9	9
	Runner System	3	9	9	27	9				9				9		9	9

		Weight														
	Gating	3	9	9	27	9				9				9		9	9
	Ejection	3	9	9	27	9				9				9		9	9
	Mold Component Life	5	15	15	45	15				15			15	15		45	15
		0															
Communication	Repair Estimate	4	0		12		12	36	4	12				12	12	12	12
	Repair Completion	5	0				15	15	5	5				15	15	5	
	Work Performed	3	0						3	3							9
		0															
Effective Tooling	Hot Runner	3	9	9	9	9				3							
	Cold Runner	3	9	0	9	9				3							
	Optimum Cycle Time	5	0	15	45	45				15						15	
	Ease of Setup	4	0														
	No Water Leak	5	0	15	15	15				15							
Technical Importance	**Absolute**		198	141	402	171	131	141	72	233	60	90	35	176	27	249	128

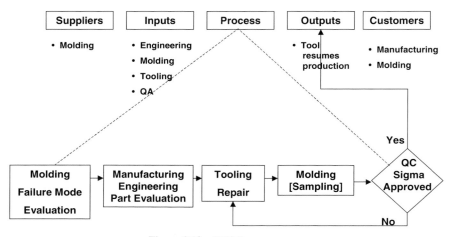

Figure 9.13   SIPOC process map.

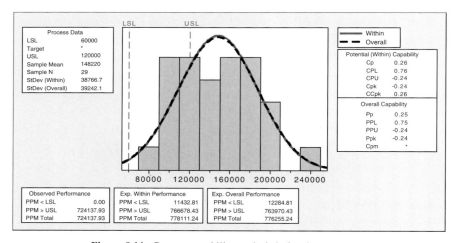

Figure 9.14   Process capability analysis before improvement.

is equal to nine. Consequently, the relationship matrix score is equal to 45, which is the product of five and nine. This is recorded in Table 9.2.

Therefore, the absolute score for *Work experience* = sum of all relationship matrix scores for all the *whats* in relationship to *hows* with *work experience*. For instance, from Table 9.2:

$$\text{Absolute score} = \sum_{i=1}^{n} x_i$$

$$\text{Absolute score for work experience} = \sum_{i=1}^{17} x_i = 45 + 45 + 6 + 6 + 45 + 15$$

$$+ 15 + 27 + 27 + 27 + 27 + 27 + 45 + 12 + 9 + 9 + 15 = 402$$

As evident from Table 9.2, the work experience score is the highest and needs improvement. It means that operators require more training to improve the process in preventing the tool damages.

9.2.3 PHASE 2: ANALYSIS OF COLLECTED DATA

The root cause analysis was mapped using the fishbone diagram in Figure 9.15. The results of for top ten tool repair expenses in damaged cores and cavities as a function of mold numbers and machine numbers are illustrated in Figures 9.16 through 9.20 using box plot, multi-vari analysis, and Pareto charts. These analyses gave us enough information to continue with the next step of the improvement phase.

9.2.4 PHASE 3: THE PROCESS OF IMPROVEMENT

A design of experiment (DOE) was carried out to identify which input factor has the highest effect in the output of the process. Three top variables (molding machine type, operator experience, and mold steel type) were selected for the design. The design was created and analyzed using the computer software Minitab.

Furthermore, results from analysis of QFD (Table 9.2), box plot, root cause analysis, and multi-vari confirmed that molding machine, operator experience, and steel type were the most dominant factors in the process.

In addition, the design of experiment was expanded with the three most dominant factors (Figures 9.21 and 9.22) in obtaining the reduced models (Figure 9.23). The results obtained from DOE for main effects and interactions were plotted in Figures 9.24 and 9.25. The Failure Mode Effects Analysis (FMEA—given in Figure 9.26), as well as the risk prioritization number (RPN—Figure 9.27 and Table 9.3), were applied for this process before and after the improvement.

RPN analysis took all the possible failure modes and effects into account.

First, the highest RPN numbers were selected in reducing the failures to make the biggest impact in the process improvement. Then lower RPN numbers were used in eliminating the failures.

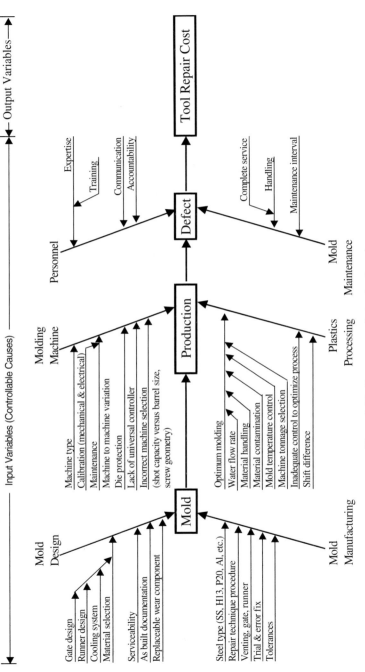

Figure 9.15 Root cause analysis of tool repair cost.

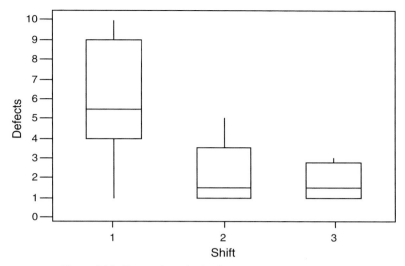

Figure 9.16   Process box plot for first, second, and third shifts.

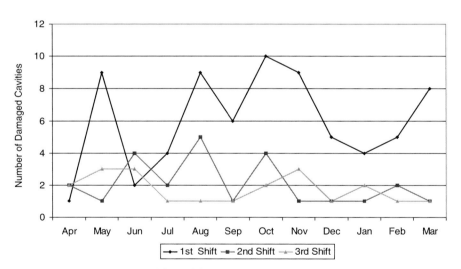

Figure 9.17   Multi-vari analysis.

Damaged Core/Cavity Cost Function of Mold

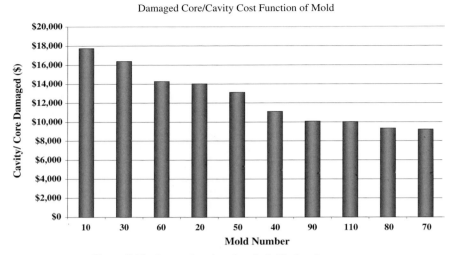

Figure 9.18 Pareto chart based on individual tool expenses.

Damaged Core or Cavity Function of Machine

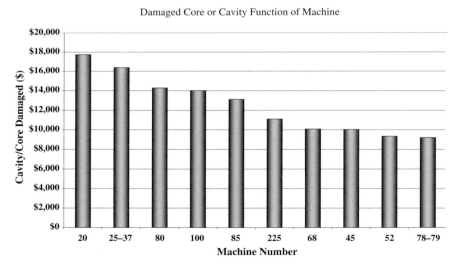

Figure 9.19 Pareto chart based on machine-related individual tool expenses.

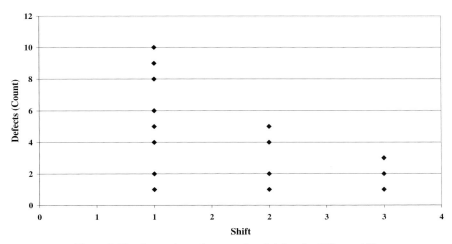

Figure 9.20   Comparison of scatter plot of defect for different shift.

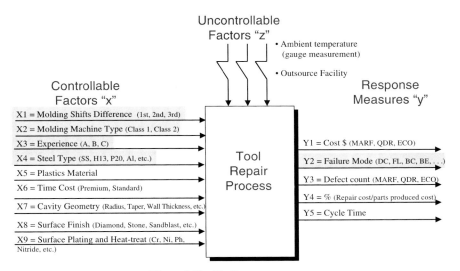

Figure 9.21   Tooling process map.

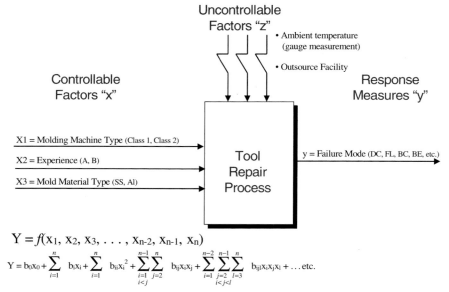

$$Y = f(x_1, x_2, x_3, \ldots, x_{n-2}, x_{n-1}, x_n)$$

$$Y = b_0 x_0 + \sum_{i=1}^{n} b_i x_i + \sum_{i=1}^{n} b_{ii} x_i^2 + \sum_{\substack{i=1 \\ i<j}}^{n-1} \sum_{j=2}^{n} b_{ij} x_i x_j + \sum_{\substack{i=1 \\ i<j<l}}^{n-2} \sum_{j=2}^{n-1} \sum_{l=3}^{n} b_{ijl} x_i x_j x_l + \ldots \text{etc.}$$

Figure 9.22   Process map for selected variables.

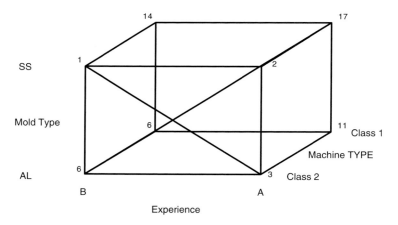

$$Y = b_0 + b_1 x_1 + b_2 x_2 + b_3 x_3 + b_{12} x_1 x_2 + b_{13} x_1 x_3 + b_{23} x_2 x_3 + b_{123} x_1 x_2 x_3$$

$$Y = 7.50 + 0.75 x_1 + 1 x_2 + 4.50 x_3 + 0.25 x_1 x_2 + 1.25 x_1 x_3 + 2.50 x_2 x_3 - 0.75 \ x_1 x_2 x_3$$

After screening the design, the reduced model comes to:

$$Y = b_0 + b_1 x_1 + b_2 x_2 + b_3 x_3 + b_{13} x_1 x_3 + b_{23} x_2 x_3$$
$$Y = 7.50 + 0.75 x_1 + 1 x_2 + 4.50 x_3 + 1.25 x_1 x_3 + 2.50 x_2 x_3$$

Figure 9.23   Design of experiment reduced models.

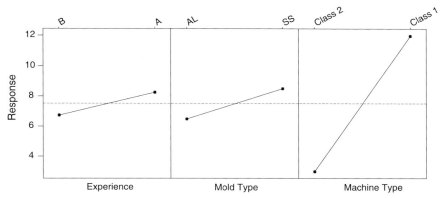

Figure 9.24 Design of experiment result main effects.

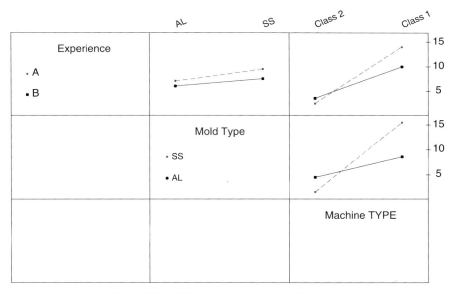

Figure 9.25 DOE interaction outcome of response.

9.2.5 PHASE 4: THE PROCESS OF CONTROL AND MAINTENANCE

Finally, additional analysis for continuous improvement purposes, such as a trend deployment flowchart and control charts $\bar{X}$ and R (Figures 9.28 and 9.29)

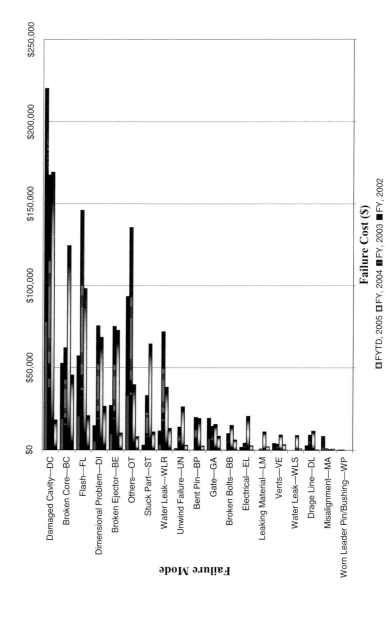

Figure 9.26 Failure mode effects cost analysis (FMEA) tool repair (years 2002–2005).

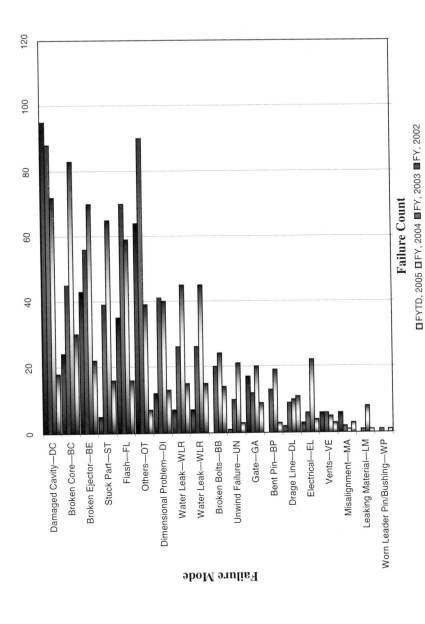

Figure 9.27 Risk prioritization number (RPN) for tool repair (years, 2002–2005).

Table 9.3
Risk Prioritization Number (RPN)
Failure Mode and Effects Analysis Worksheet

Items and function	Potential failure mode	Failure code	Potential effect(s) of failure	Severity	Potential cause(s) of failure	Occurrence	Current controls	Detection	RPN	Recommended action	Responsibility and target completion	Action taken	Severity	Occurrence	Detection	RPN
					FMEA process								Action results			
Mold Cavity	Damaged Cavity	DC	Production	10	Die Protection/Cold water	9	20% inspected/ongoing	4	360	Examine machine process/design/Tool	Tracker Team/ Marco, John	Improve Eng. design & mold	10	8	3	240
	Broken Damaged Core	BC	Production	10	Die Protection/ Cold water	8	20% inspected/ongoing	3	240	Examine machine process/design/Tool	Tracker Team/ Salman, James	Improve Eng. design & mold	10	5	3	150
	Flash	FL	Dimensional	9	Clamp Tonnage, Mold damage	8	25% inspected/ongoing	4	288	Improve repair	Molding-tooling/ Marco-John	Training	8	5	4	160
	Stuck Part	ST	Production	9	Temperature-Ejector pin	8		3	216	Improve process/repair	Molding/ Marco	Training	8	6	3	144
	Broken Ejector	BE	Production	8	Binding, temperature	7		4	224	Lubrication/Material	Tooling/ John	Lubrication/ Material	8	6	4	192
	Dimensional Problem	DI	Part Function	8	Water flow	8		3	192	Wider tolerance	Eng./James	Design/ Cooling system	8	5	2	80
	Water Leak	WL	Production	9	Damaged gasket seal, Cracked Cavity/Core	8	70% inspected/ongoing	4	288	Pressure testing	Molding-Tooling/ Marco, John	Oil ring lubrication, redesigns pockets	8	6	4	192

	Item	Code	Function	Sev	Failure Cause	Occ	Current Controls	Det	RPN	Recommended Action	Responsibility	Action Taken	Sev	Occ	Det	RPN
	Vents	VE	Plastics Flow	8		6	70% inspected/ ongoing	2	96	Optimum vent design before mold mfg	Tooling/ John	Optimum vent design before mold repair	8	5	2	80
	Gate	GA	Plastics Flow	5	Cleaning tools	6	50% inspected/ ongoing	4	120	Optimum vent design before mold mfg	Tooling/ John	Optimum vent design before mold repair	5	5	3	75
	Drag Line	DL	Cosmetic	5	Hot/Cold melt temperature	5		2	50	None	Molding/ Marco		5	5	2	50
	Electrical	EL	Production	4	Command signal not repeatable	5		3	60	None	Maintenance/ Jim		4	5	3	60
	Unwind Failure	UN	Production	8	Hydraulic motor failure	8	90% inspected/ ongoing	3	192	Control pressure	Maintenance/ Jim	Installed pressure controller	8	2	5	80
	Broken Bolt	BB	Mold Function	8		8	70% inspected/ ongoing	5	320	None	Tooling/ John		8	8	5	320
	Leaking Material	LM	Dimensional	4		8		5	160	None			4	8	5	160
	Bent Pin	BP	Part Function	8		8		5	320	Modification	Tooling/ John	Modification	8	6	5	240
	Worn Leader Pin/Bushing	WP	Part Function	5	Platen misalignment	5	100% inspected	5	125	None			5	5	5	125
Tool	Mold Damaged During Maintenance	MD	Down Time	2	Handling	2	100% inspected	9	36	Training	Tooling/ John	Training	2	1	5	10
	Misalignment	MA		2		2	50% inspected/ ongoing	8	32	None	Tooling/ John		2	2	8	32
	Flower Pot	FP	Part Function	7		8	30% inspected/ ongoing	5	280	Instal vision system	Tooling/ John	None	8	7	5	280
	Other	OT		8		5	ongoing inspection	5	200		Tracker team		8	4	5	160

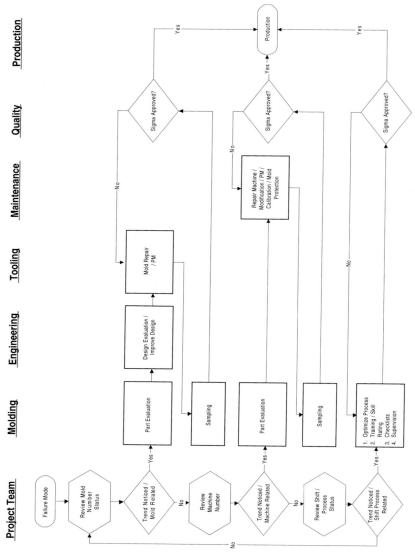

Figure 9.28 Deployment flowchart of trend analysis.

202

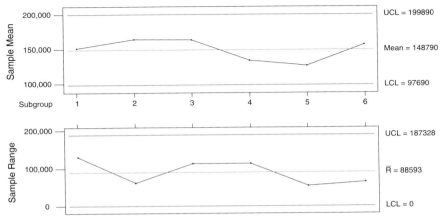

Figure 9.29 Control chart for tool repair cost.

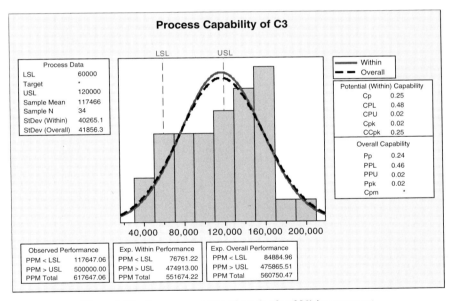

Figure 9.30 Process capability of repair after 20% improvement.

and process capability (Figure 9.30), were implemented. After a few months, results confirmed the continuous improvement and financial benefit of this project.

Conclusion: The total repair cost was reduced to 1.558×10^6 in the first year. The project still maintains the lower cost, and the cost continues to decline.

Chapter 10

Six Sigma Master Black Belt Level Case Study

10.1 CASE STUDY: DEFECT REDUCTION IN INJECTION MOLDING A MULTIFACTOR LEAN CENTRAL COMPOSITE DESIGN APPROACH

10.1.1 SCOPE OF INJECTION MOLDED PARTS

A great deal of emphasis has been placed on the repeatability of the injection molding operation (Figures 10.1 through 10.4, for a typical injection molding machine refer to Figure 6.7) through the use of feedback control techniques. However, it is the molded parts whose quality determines performance, sale, and reputation. There are inherent changes that occur in the processing conditions with time that might affect such properties. Variation in the material composition, molecular weight, or molecular weight distribution from lot to lot cannot be compensated for by maintaining machine repeatability. Neither can variations in the proportion or quality of regrind in the feed. The thermal degradation of hydraulic fluid, changes in the heater-band characteristics, and wear in the barrel or screw could also call into question the ability of the press (injection molding machine) to make equivalent parts through the life of the machine.

10.1.2 SCOPE OF STUDY

An experimental study of injection molding was conducted by using a three-factor composite design with the independent variables—injection pressure

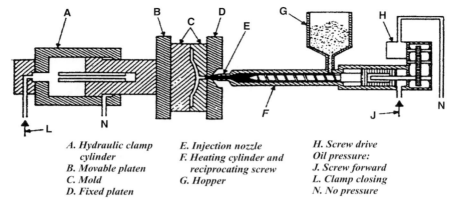

Figure 10.1   Injection molding cycle—melt ready for injection.

A. Hydraulic clamp
  cylinder
B. Movable platen
C. Mold
D. Fixed platen

E. Injection nozzle
F. Heating cylinder and
  reciprocating screw
G. Hopper

H. Screw drive
Oil pressure:
J. Screw forward
L. Clamp closing
N. No pressure

(Hi = 14,000 psi, Low = 6,000 psi); injection velocity (Hi = 80 mm/sec, Low = 40 mm/sec); and plasticating screw speed (Hi = 200 rpm, Low = 100 rpm)—to the thickness, weight, and strengths of the molded parts.

One advantage of this choice of variables is that the responses to their change do not have substantial time lags. In contrast, if one chooses to adjust a temperature, many cycles (and parts) would be required to stabilize the processing conditions before data collection could be reestablished.

The "part quality" (Y) consisted of two nondestructive properties, weight, and thickness, and one destructive property, either tensile modulus or impact strength, depending on the portion of the molded part being described. The process under study can be represented as in Figure 10.5. A cause-and-effect diagram, as shown in Figure 10.6, was used in selection of design variables.

10.2 COMPOSITE DESIGN METHODOLOGY

Composite design is the most well-known design for estimating the coefficients up to the second-order model, which is also called the central composite design. This method was originally introduced by Box and Wilson. (For further information, refer to any experimental design and analysis texts.) Here, the central composite design consists of the following:

1. Two-level 2^3 factorial, where the factor levels are coded values x_1, x_2, x_3 = ±1, ±1, ±1.
2. Six axial or star points in coded values (+2, 0, 0), (0, 0, +2), where ±2 is the distance from the center of the design on the axis of each design.
3. Center point is coded values $x_1, x_2, x_3 = 0, 0, 0$.

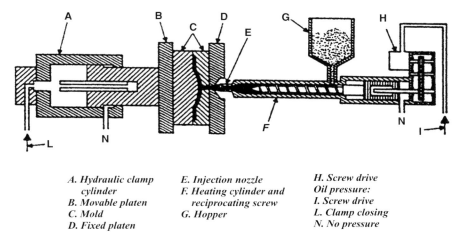

A. Hydraulic clamp cylinder	E. Injection nozzle	H. Screw drive
B. Movable platen	F. Heating cylinder and reciprocating screw	Oil pressure:
C. Mold	G. Hopper	I. Screw drive
D. Fixed platen		L. Clamp closing
		N. No pressure

Figure 10.2 Injection molding cycle—melt being injected.

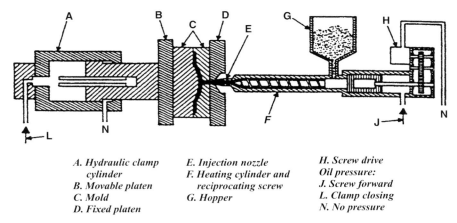

A. Hydraulic clamp cylinder	E. Injection nozzle	H. Screw drive
B. Movable platen	F. Heating cylinder and reciprocating screw	Oil pressure:
C. Mold	G. Hopper	J. Screw forward
D. Fixed platen		L. Clamp closing
		N. No pressure

Figure 10.3 Injection molding cycle—mold closed for cooling (melt being plasticized).

The total number of design points (trials) can be calculated by the number of factorial points (2^3) + axial points (2 times 3) + center point (1) equal to 15. Then the center point was replicated five times to get an estimate of "experimental error." Table 10.1 shows the coded variables for this design. The coded variables 0, ± 1, ± 2 are factor levels for the experimental design, calculated using Equation 10.1.

Table 10.1

**A Three-Factor Composite Design (Coded Variables) of Injection
Pressure, Injection Velocity, and Plastication Screw Speed**

Trial number	Injection pressure (x_1)	Injection velocity (x_2)	Screw speed (x_3)
1 (Corner)	+1	+1	+1
2 (Corner)	+1	+1	−1
3 (Corner)	+1	−1	+1
4 (Corner)	+1	−1	−1
5 (Corner)	−1	+1	+1
6 (Corner)	−1	+1	−1
7 (Corner)	−1	−1	+1
8 (Corner)	−1	−1	−1
9 (Star)	+2	0	0
10 (Star)	−2	0	0
11 (Star)	0	+2	0
12 (Star)	0	−2	0
13 (Star)	0	0	+2
14 (Star)	0	0	−2
15 (Center)	0	0	0
16 (Replicated)	0	0	0
17 (Replicated)	0	0	0
18 (Replicated)	0	0	0
19 (Replicated)	0	0	0
20 (Replicated)	0	0	0

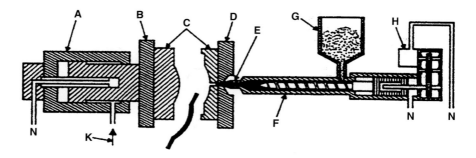

A. Hydraulic clamp
    cylinder
B. Movable platen
C. Mold

D. Fixed platen
E. Injection nozzle
F. Heating cylinder and
    reciprocating screw
G. Hopper

H. Screw drive
Oil pressure:
K. Clamp opening
N. No pressure

Figure 10.4   Injection molding machine part ejection.

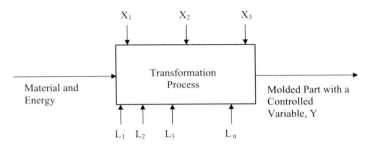

Figure 10.5   General model of the process.

$$\text{Coded} = \left(\frac{2}{High - Low} \right) Actual - 1 \qquad (10.1)$$

Rearranging 10.1 gives us Equation 10.2:

$$\text{Actual} = \left(\frac{High + Low}{2} \right) + \left(\frac{High - Low}{2} \right) \text{Coded} \qquad (10.2)$$

The modified version of Equation 10.1 in terms of pressure, velocity, and screw speed is shown in Equations 10.3–10.6.

$$CV = \frac{Inj - Inj(0)}{\Delta Inj} \qquad (10.3)$$

where
CV: Coded variable
Inj: Injection "condition" at 0, ±1, ±2
Inj(0): Injection condition at the center point of the design.

For example, in the case of injection pressure x_1, injection velocity x_2, and plastication screw speed x_3, we have, respectively,

$$x_1 = \left(P_{Inj} - 10{,}000 \right) / 4{,}000 \qquad (10.4)$$

where P_{Inj} is the uncoded injection pressure in psi,

$$x_2 = \left(V_{Inj} - 60 \right) / 20 \qquad (10.5)$$

where V_{Inj} is the uncoded velocity in mm/s, and

$$x_3 = \left(SS - 150 \right) / 50 \qquad (10.6)$$

where SS is uncoded plastication screw speed in rpm.

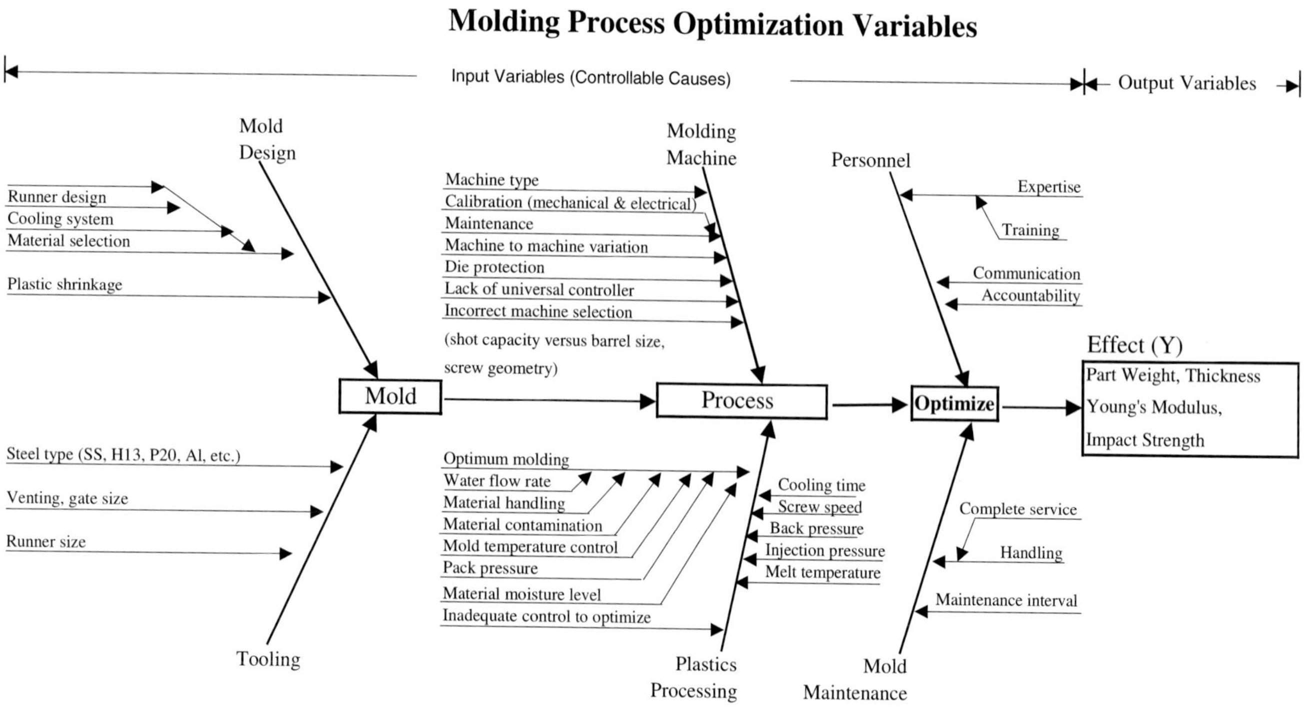

Molding Process Optimization Variables

Input Variables (Controllable Causes) → ← Output Variables →

Effect (Y)
Part Weight, Thickness
Young's Modulus,
Impact Strength

Personnel
Expertise
Training
Communication
Accountability

Optimize

Mold Maintenance
Complete service
Handling
Maintenance interval

Molding Machine
Machine type
Calibration (mechanical & electrical)
Maintenance
Machine to machine variation
Die protection
Lack of universal controller
Incorrect machine selection
(shot capacity versus barrel size, screw geometry)

Process

Plastics Processing
Cooling time
Screw speed
Back pressure
Injection pressure
Melt temperature
Optimum molding
Water flow rate
Material handling
Material contamination
Mold temperature control
Pack pressure
Material moisture level
Inadequate control to optimize

Mold Design
Runner design
Cooling system
Material selection
Plastic shrinkage

Mold

Tooling
Steel type (SS, H13, P20, Al, etc.)
Venting, gate size
Runner size

Figure 10.6 Cause-and-effect diagram.

210

The resulting order of the experimental design points is shown in Figure 10.7. Each batch included 50 parts, which were numbered in their sequential order of manufacture, and then each test was done on the last 25 parts, numbers 26 through 50.

The coded values of the corners for the cube ([±1, 1, 1], [1, ±1, 1], and [1, 1, ±1]) and the star or axial points for the XYZ axis ([±2, 0, 0], [0, ±2, 0], and [0, 0, ±2]) listed in Table 10.1 represents Figure 10.7. The process model (Equation 10.7) from the design chart is

$$Y = b_0 + b_1 x_1 + b_2 x_2 + b_3 x_3 \qquad \text{Linear terms}$$
$$+ b_{11} x_1^2 + b_{22} x_2^2 + b_{33} x_3^2 \qquad \text{Quadratic effects} \qquad (10.7)$$
$$+ b_{12} x_1 x_2 + b_{13} x_1 x_3 + b_{23} x_2 x_3 \qquad \text{Binary interaction effects}$$

Experimental

A composite experimental design as discussed above, with three operating conditions (independent variables), as shown in Figure 10.7 and Table 10.1— injection pressure x_1, injection velocity x_2, and plasticating screw speed x_3—each at five levels, was used to mold ASTM impact-strength square plaques (51 mm × 51 mm, nominally 3-mm thick) and double tensile test bars (each with a 15-mm-long rectangular neck, nominally 3 mm × 3 mm), using high-density polyethylene

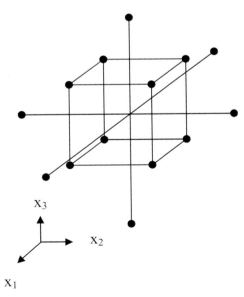

Figure 10.7   Three-factor nonorthogonal composite design.

(HDPE), as shown in Figures 10.8 and 10.9. At each of the 14 noncenter experimental conditions, 25 preliminary shots were made, followed by 25 property evaluations. At the center-of-the-design condition, the 25 preliminary shots were followed by six replications of 25, or 150 shots; these were used to obtain estimates of the standard deviations in part properties.

10.3 MODELING

Although model equations were derived for each of the properties measured—plaque weight and thickness, tensile bar weight and thickness, plaque impact strength, and tensile bar Young's modulus (at low speed and high speed, this is shown in Figures 10.10 and 10.11, while Figure 10.12 shows how Young's modulus can be determined by using the stress-strain ratio which is a modulus [stiffness] of elasticity defined as the ratio of applied stresses [tensile strength] to strain [elongation] at yield.

$$\text{Young's Modulus} = \text{Modulus of elasticity} = E = \frac{Stress}{Strain} = \frac{\sigma}{\varepsilon}$$

E is Young's modulus in psi/in, σ is applied tensile stress [force/area] in pounds per square inch and ε is the elongation [strain] in inches.)—the simulation utilized only the plaque weight (the plaque proper is formed from a fan gate; the "part" was broken from the runner at the fan-gate entrance, rather than at the gate-plaque interface.) The weight (g) of this "part" was measured to the nearest 0.1 mg.

The plaque weight was found to be given by Equation 10.8.

$$Y_{pw} = 8.8321 - 0.0064X_1^2 + 0.0090X_2^2 + 0.0058X_3^2 \tag{10.8}$$

This equation of the model is known as the canonical form because the original coordinate axes have been translated and rotated so as to "remove" the linear and interaction terms. It gives the dependent variable, plaque weight, in the measured units (g) as a function of the canonical coordinated axes, X. Equations 10.9–10.11 are for these axes.

$$X_1 = -1.0000(x_1 - x_{1s}) + 0.0036(x_2 - x_{2s}) + 0.0075(x_3 - x_{3s}) \tag{10.9}$$

$$X_2 = -0.5402(x_1 - x_{1s}) + 0.5253(x_2 - x_{2s}) + 0.6575(x_3 - x_{3s}) \tag{10.10}$$

$$X_3 = -0.2509(x_1 - x_{1s}) + 0.8959(x_2 - x_{2s}) + 0.3666(x_3 - x_{3s}) \tag{10.11}$$

Here, the x_i are the independent variables, in coded form, and the x_{is} are the values of these coded variables at the "stationary" point of the response surface. The coded variables and their stationary point values are given by Equations 10.4–10.6; see Equations 10.12–10.14.

Figure 10.8   Part as molded.

Figure 10.9   Double tensile bar and plaque separated from sprue and runners.

$$x_1 = \left(P_{inj} - 10000\right)/4000 \text{ with } x_{1s} = 0.0829 \tag{10.12}$$

$$x_2 = \left(V_{inj} - 60\right)/20 \text{ with } x_{2s} = 0.5060 \tag{10.13}$$

$$x_3 = \left(SS - 150\right)/50 \text{ with } x_{3s} = -2.0272 \tag{10.14}$$

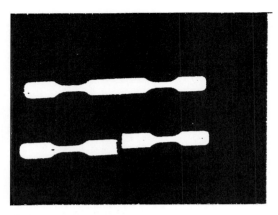

Figure 10.10   Double tensile bar as molded and cut into two (single) bars.

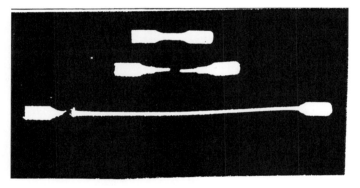

Figure 10.11   Tensile bar before testing, after high-speed test, and after low-speed test.

The numerical values in the defining equations are those used by the injection molding machine: the "press." The 10,000 and 4,000 are psi, equivalent to 68.98 MPa and 27.59 MPa, respectively. The 60 and 20 are mm/s. The 150 and 50 are rpm, equivalent to 15.7 rad/s and 5.24 rad/s, respectively.

The experimental design used values of the coded variables equal to −2, −1, 0, 1, and 2. Thus, the injection pressures (Tables 10.1 and 10.2) used were 2,000, 6,000, 10,000, 14,000, and 18,000 psi (13.80, 27.59, 68.98, 96.57, and 124.16 Mpa, respectively). The injection velocities used were 20, 40, 60, 80, and 100 mm/s. The plasticating screw speeds were 50, 100, 150, 200, and 250 rpm (5.236, 10.47, 15.71, 20.94, and 26.18 rad/s, respectively).

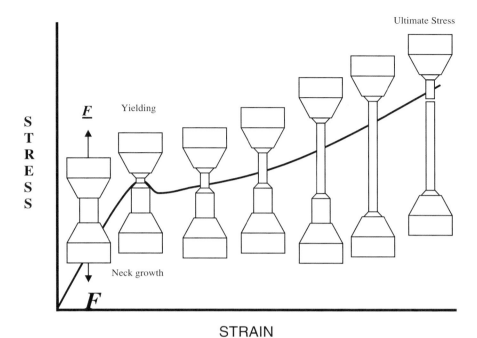

STRAIN

Figure 10.12 Illustration of tensile stress-strain as force is applied gradually (low speed) until break point.

Table 10.2 displays the uncoded variables of Table 10.1, and Table 10.3 presents the uncoded percentage variables as they appear on the injection molding machine control panel.

Table 10.4 illustrates the parameters that held constant during the experiment.

10.4 SIMULATION

In order to compare the effect of a controller, the simulation, using MODSIM II (a short form of "modular block structure high-level simulation programming language"), was run first in the open-loop mode and then in the closed-loop mode for each control strategy. Two types of controllers were used: a "conventional" integral-only controller and a reinitializing CUSUM (statistical process control algorithm) SPC controller.

The average experimental value of the plaque weights at the center-of-the-design conditions (8.8729 g) was used as the set point. The weights of the "middle"

Table 10.2

A Three-Factor Composite Design (Uncoded Variables) of Injection Pressure, Injection Velocity, and Plastication Screw Speed

Trial number	Injection pressure (Plastic pressure) psi	Injection velocity mm/s	Screw speed rpm
Maximum Value	18,420	100	310
1 (Corner)	14,000	80	200
2 (Corner)	14,000	80	100
3 (Corner)	14,000	40	200
4 (Corner)	14,000	40	100
5 (Corner)	6,000	80	200
6 (Corner)	6,000	40	200
8 (Corner)	6,000	40	100
9 (Star)	18,000	60	150
10 (Star)	18,000	60	150
11 (Star)	10,000	99	150
12 (Star)	10,000	20	150
13 (Star)	10,000	60	250
14 (Star)	10,000	60	78
15 (Center)	10,000	60	150
16 (Replicated)	10,000	60	150
17 (Replicated)	10,000	60	150
18 (Replicated)	10,000	60	150
19 (Replicated)	10,000	60	150
20 (Replicated)	10,000	60	150

99 (i.e., ordinal numbered parts 26 through 124) of the 150 molded at these conditions were plotted on cumulative probability paper to verify that the distribution was normal. The standard deviation of this distribution was 0.0108.

Open-Loop Performance

In the open-loop mode, the independent variables were set at their center-of-design coded values, $x_{ic} = 0$, a shot was "made" and the weight of the plaque calculated, and the random normal-distribution generator (again, using the experimental value of the standard deviation) added (or subtracted) a value to (or from) this weight. The result was the simulated value of the plaque weight.

Now this value was subtracted from the set point, the difference being the error signal sent to the controller (Figure 10.13). The action of the controller changed the operating condition(s), using the modeling equations, and a subsequent shot was made. The resulting calculated plaque weight was adjusted

Table 10.3

A Three-Factor Composite Design (Uncoded Percentage Variables) of Injection Pressure, Injection Velocity, and Plastication Screw Speed

Trial number	Injection pressure (Plastic pressure) psi	Injection velocity mm/s	Screw speed rpm
Maximum Value	100	100	100
1 (Corner)	76	80	64.52
2 (Corner)	76	80	32.26
3 (Corner)	76	40	64.52
4 (Corner)	76	40	32.26
5 (Corner)	32.57	80	64.52
6 (Corner)	32.57	80	32.26
7 (Corner)	32.57	40	64.52
8 (Corner)	32.57	40	32.26
9 (Star)	97.72	60	48.39
10 (Star)	10.89	60	48.39
11 (Star)	54.29	99	48.39
12 (Star)	54.29	20	48.39
13 (Star)	54.29	60	80.65
14 (Star)	54.29	60	77.50
15 (Center)	54.29	60	48.39
16 (Replicated)	54.29	60	48.39
17 (Replicated)	54.29	60	48.39
18 (Replicated)	54.29	60	48.39
19 (Replicated)	54.29	60	48.39
20 (Replicated)	54.29	60	48.39

Table 10.4

Parameters Held Constant in the Molding Optimization Study

Molding Machine Specification	Max. clamp force tonnage = 15 ton
	Part ejection pressure = 40% psi
	Back pressure = 00% psi
Material	HDPE
Resin Specification	Drying = 200°F
	Melt temperature = 248°F
	Front zone = 450°F
	Rear zone = 350°F
	Mold (Stationary side) = 120°F
	Mold (Moving side) = 120°F
Time	Injection = 10 sec
	Cooling = 12 sec
	Ejection = 2 sec

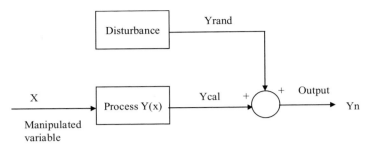

Figure 10.13   Open-loop process model.

for random disturbances, and the simulation sequence repeated. After 150 shots, the mean value and standard deviation of the distribution of the plaque weights were calculated.

The resulting mean values were close to those of the experimental data. If the value of the standard deviation for closed-loop simulation was less than that of the experimental data, it was inferred that if such a controller had been implemented, the performance of the process could have been improved. If not, such a control strategy was regarded as unacceptable.

One Variable Integral

When only one operating condition was allowed to change, the entire error signal was used to adjust that variable before the next shot was simulated. A finite sum was used to approximate the integral where the sampling period, Δt, was one shot. A single tuning constant was used to determine the magnitude of the adjustment. (The details are in the Appendix. Note that this adjustment does not utilize the modeling equations.)

In no case with the integral controller were either the open-loop or closed-loop standard deviations as small as in the experimental value. This is understandable, since the modeling equations showed that these variables had interaction terms that affected the plaque weight. The numerical results are shown in Table 10.5.

One Variable CUSUM

When only one operating condition was allowed to change (Figure 10.14), the entire error signal was used to adjust that variable before the next shot was simulated. The partial derivative of the dependent variable with respect to the operating condition was used to determine the magnitude of this adjustment. (The details are in the Appendix.)

Table 10.5
Results of Integral Control

Manipulated variable	Mode	Mean	Standard deviation
x_1	Open	8.8651	0.01134
	Closed	8.8686	0.01160
x_2	Open	8.8652	0.01122
	Closed	8.8636	0.01097
x_3	Open	8.8642	0.01134
	Closed	8.8683	0.01092

Table 10.6
Results of One Variable CUSUM Control

Manipulated variable	Mode	Mean	Standard deviation
x_1	Open	8.8651	0.01134
	Closed	8.8626	0.01297
x_2	Open	8.8652	0.01122
	Closed	8.8737	0.01209
x_3	Open	8.8642	0.01134
	Closed	8.8754	0.01436

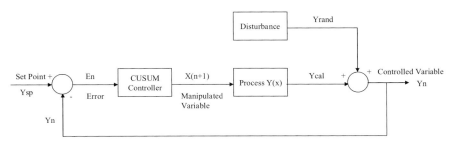

Figure 10.14   Closed-loop control model with one variable.

In no case, with the CUSUM controller, was either the open-loop or closed-loop standard deviation as small as the experimental value. Again, this is understandable because the interaction terms have been excluded. The numerical results are shown in Table 10.6.

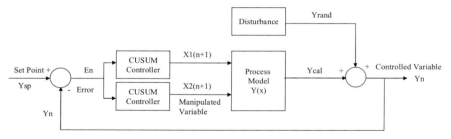

Figure 10.15   Closed-loop control model with two variables.

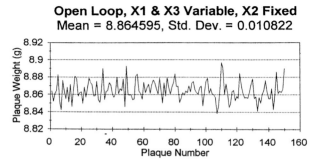

Figure 10.16a   Plaque weight in open loop.

Two Variables CUSUM

When two operating conditions were allowed to change, the error signal was split in order to adjust those variables before the next shot was simulated (Figure 10.15). Here, a single weighting factor was used to "tune" this split in order to "minimize" the standard deviation achievable. Again, the partial derivatives of the dependent variable, with respect to the operating conditions, were used to determine the magnitudes of these adjustments. (See Appendix for details.)

In one such case, with the CUSUM controller, the closed-loop standard deviation was smaller than the experimental value, as shown in Figures 10.16a and b. In another case, that for the open-loop was even smaller—that is, less than the experimental value, the corresponding closed-loop value, and that of the preceding improvement. Other than randomness, there is no explanation for this. Of course, the authors have not demonstrated that the smaller closed-loop value found is, indeed, the smallest value achievable.

Such an improvement is understandable, since the modeling equations showed that these variables had interaction terms that affected the plaque weight. Adjust-

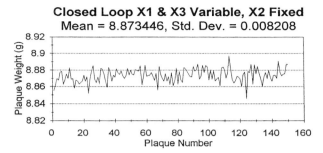

Figure 10.16b   Plaque weight in closed loop, using CUSUM controller with tuning constant 0.74855.

Table 10.7
Results of Two Variables CUSUM Control

Manipulated variable	Mode	Mean	Standard deviation
x_1, x_2	Open	8.8655	0.01133
	Closed	8.8755	0.01557
x_1, x_3	Open	8.8646	0.01082
	Closed	8.8734	0.00821*
x_2, x_3	Open	8.8644	0.00805**
	Closed	8.8759	0.01046

*This case demonstrates the viability of the SPC control strategy.
**In this case, the open-loop simulation had a smaller standard deviation than the experimental value.

ing two variables to control the process seems necessary to improve the performance. The numerical results are shown in Table 10.7.

Three Variables CUSUM

If three operating conditions are allowed to change, the error signal must be split in order to adjust those variables before the next shot would be simulated. Here, two weighting factors would be required to "tune" this split in order to "minimize" the standard deviation achievable. Again, the partial derivatives of the dependent variable, with respect to the operating conditions, would be used to determine the magnitudes of these adjustments. (The details are in the Appendix.) No simulation runs were made for this case because of the complexities involved.

10.5 CONCLUSION

It is considered a vindication of the proposed strategy to have found one sub-stantiating case for which it works. For those manufacturers of large, valuable parts, whose quality must be assured, the implementation of SPC control strategy to reduce the variation in properties part-to-part is recommended. Such an approach requires a reliable model derived from a suitable experimental design.

BIBLIOGRAPHY

1. Benbow, W. Donald, W. Roger Berge, K. Ahmad Elshennawy, and F. H. Walker, *The Certified Quality Engineer Handbook*. ASQ Quality Press, Milwaukee, WI (2002).
2. Breyfogle III, F. W., *Implementing Six Sigma, Smarter Solutions Using Statistical Methods*. John Wiley, New York (1999).
3. Boyle, K., *Six Sigma Black Belt Lecture Notes*. The University of California San Diego *Six Sigma* Black Belt Program (April 2003).
4. Box, G. E. P., and K. B. Wilson, "On the Experimental Attainment of Optimum Conditions," *Journal of the Royal Statistical Society*, B, 13, pp. 1–45 (1951).
5. Dennis, P., *Lean Production Simplified*. Productivity Press, New York (2002).
6. Deshpande, P. B., *Emerging Technologies and Six Sigma*. Hydrocarbon Processing, April (1998).
7. Ehrlich, B. H., *Transactional Six Sigma and Lean Servicing*. CRC Press, New York (2002).
8. Ellis, K., *Mastering Six Sigma*. Training (December 2001).
9. www.epa.gov/lean/thinking/kanban.htm (2005).
10. Microsoft, *"Excel"* (2000).
11. http://www.fourmilab.ch/rpkp/experiments/analysis/chicalc.html.
12. Michael, George L. *Lean Six Sigma*. McGraw-Hill, New York (2002).
13. Griffith, G. K., *The Quality Technician's Handbook*, 5th ed. Prentice Hall, Columbus, OH, (2003).
14. Gross, J. M., *A Road Map to Six Sigma Quality*. Quality Progress (November 2001).
15. Harry, M. J., *The Vision of Six Sigma*. Tri Star Publishing, Phoenix, AZ (1997).
16. Harry, M. J., and J. R. Lawson, *Six Sigma Producibility Analysis and Process Characterization*. Addison-Wesley Publishing, New York (1992).
17. Hunter Industries, "Mission, Vision, Values." San Marcos, CA (2003).
18. Kvanli, A. H., R. J. Pavur, and C. S. Guynes, *Introduction to Business Statistics*, 5th ed. South-Western College Publishing Mason (2000).
19. Laraia, C. A., E. P. Moody, and W. R. Hall, *The Kaizen Blitz*. John Wiley, New York (1999).
20. Montgomery, D. C., *Design and Analysis of Experiments*, 3rd ed. John Wiley, New York (1991).
21. Nakajima, S., *Introduction to Total Production Maintenance*. Productivity Press, Cambridge, MA (1988).
22. *NIST/SEMATECH e-Handbook of Statistical Methods*. http://www.itl.nist.gov/div898/handbook/ (2005).
23. Rutter, Bryce, G., "Why Ergonomics?" *Plastics Engineering* (May 1999).
24. Schmidt, S. R., and R. G. Launsby, *Understanding Industrial Designed Experiments*, 4th ed. Air Academy Press, Colorado Spring, CO (1994).
25. Smith, B., *Lean and Six Sigma—A One-Two Punch*. Quality Progress (April 2003).
26. Snee, R. D., and R. W. Hoerl, *Leading Six Sigma*. FT Prentice Hall, New York (2003).
27. Taghizadegan, S., *Injection Molding: Experimental Modelling, Simulation, and Statistical Process Control*. Ph.D. Dissertation, University of Louisville, Louisville, KY (1996).
28. Taghizadegan, S., and D. O. Harper, *Statistical Process Control of Injection Molding Simulation Based on an Experimental Study*. SPE ANTEC Technical Papers, 42, 598–602 (1996).

Appendix

Statistical Tables Used for Lean *Six Sigma*

Appendix I

Highlights of Symbols and Abbreviations

ANOVA	Analysis of Variance
b_0	y-intercept in the prediction equation
b_1	The coefficient of linear effect of variable in a regression model
b_{11}	The coefficient of quadratic term for the first variable in a regression model
C_p	Process capability (assumes process centered on the target)
C_{pk}	Process capability index (does not assume process is centered on the target)
DOE	Design of experiment
df	Degrees of freedom
f	Function (used to describe a function)
$f(x)$	Function of x
P_p	Process performance
P_{pk}	Process performance index
s	Sample standard deviation
USL	Upper specification limit
LSL	Lower specification limit
z	A standardized value
μ	Mean
σ	Standard deviation of population
$\int$	Integral of
$\int_a^b$	Integral between the limits a and b
Σx	The sum of all of the values in a set of values
$\sum_{j}^{n} x_j$	The sum of values from j = 1 to n, which is equivalent to Σx only when n equals total number of values
$\lvert x \rvert$	Absolute value of x; if x is a negative number, the sign is ignored and the outcome will be a positive value
log	Common logarithm
ln	Natural logarithm
E	Exponential notation. Example $1.67E + 2$ is the same as 16.7×10^2

Appendix II

Chapter 10 Case Study Extended Equations

One Variable Integral

For the integral control of a single variable,

$$x_i = \tau \int_0 \varepsilon \, dt \approx \tau \sum_j \varepsilon^{(j)} \Delta t, \ j = 0, 1, \ldots n$$

$$\therefore x_i^{(n+1)} = x_j^{(n)} + \tau \varepsilon^{(n)}, \text{ where } \varepsilon^{(n)} = y_{sp} - y^{(n)}.$$

One Variable CUSUM

For the CUSUM control of one variable, the modeling equation (in the non-canonical form) is used.

$$y = b_0 + b_1 x_1 + b_2 x_2 + b_3 x_3 + b_{11} x_1^2 + b_{22} x_2^2 + b_{33} x_3^2 + b_{12} x_1 x_2 + b_{13} x_1 x_3 + b_{23} x_2 x_3.$$

Therefore,

$$(\partial y / \partial x_i) = b_i + 2 b_{ii} x_1 + \Sigma_j b_{ij} x_j, \ j \neq i.$$

$$(\partial y / \partial x_j)^{(n)} \approx (y^{(n+1)} - y^{(n)} / (x_i^{(n+1)} - x_i^{(n)}).$$

$$\therefore x_i^{(n+1)} = x_i^{(n)} + (y_{sp} - y^{(n)}) / (b_i + 2 b_{ii} x_i^{(n)} + \Sigma_j b_{ij} x_j).$$

Two Variables CUSUM

For the CUSUM control of two variables, the modeling equation (in the non-canonical form) is also used.

$$dy = (\partial y / \partial x_i) dx_i + (\partial y / \partial x_j) dx_j = (b_i + b_{ij} x_j + b_{ik} x_k + b_{ii} x_j) \, dx_i.$$

$$\therefore (y_{sp} - y^{(n)}) \approx (b_i + b_{ij} x_j + b_{ik} x_k + 2 b_{ii} x_i)(x_i^{(n+1)} - x_i^{(n)})$$

$$+ (b_j + b_{ij} x_j + b_{jk} x_k + 2 b_{jj} x_j)(x_j^{(n+1)} - x_j^{(n)}),$$

or,
$$(y_{sp} - y^{(n)}) \approx c_i^{(n)}(x_i^{(n+1)} - x_i^{(n)}) + c_j^{(n)}(x_j^{(n+1)} - x_j^{(n)}).$$

Introducing a weighting parameter, w, such that $wc_i^{(n)} x_i^{(n+1)} = (1 - w)c_j^{(n)}x_j^{(n+1)}$, the controller equations become

$$x_i^{(n+1)} = (1 - w)(y_{sp} - y^{(n)} + c_i^{(n)}x_i^{(n)} + c_j^{(n)}x_j^{(n)}/c_i^{(n)}$$

and

$$x_j^{(n+1)} = w(y_{sp} - y^{(n)} + c_i^{(n)}x_i^{(n)} + c_j^{(n)}x_j^{(n)})/c_j^{(n)}.$$

Three Variables CUSUM

Using the same method as for two variables, the CUSUM control of three variables requires two weighting factors. These are defined such that

$$w_i c_i^{(n)} x_i^{(n+1)} = w_j c_j^{(n)} x_j^{(n+1)} = w_k c_k^{(n)} x_k^{(n+1)} = (1 - w_i - w_j)c_k^{(n)}x_k^{(n+1)}.$$

One of the resulting controller equations is

$$x_i^{(n+1)} = (y_{sp} - y^{(n)} + c_i^{(n)}x_i^{(n)} + c_j^{(n)}x_j^{(n)} + c_k^{(n)}x_k^{(n)}) = [(2w_i + w_j w_k)/w_j w_k]/c_i^{(n)};$$

the others are similar.

Nomenclature

English

b_i, b_{ij}	coefficients in the model equation
c_i	coefficients in the controller equations
P_{inj}	injection pressure (psi)
SS	plastication screw speed (rpm)
V_{inj}	injection velocity (mm/s)
w, w_i	weighting factors in the controller equations
x_1	coded injection pressure
x_2	coded injection velocity
x_3	coded plastication screw speed
X_i	canonical axes (i = 1, 2, 3)
x_i	independent variable (generic)
x_{ic}	coded center-of-the-design values (i = 1, 2, 3)
x_{is}	coded stationary-point-of-the-response-surface values (i = 1, 2, 3)
Y_{PW}	plaque weight (g)
y	dependent variable (generic)
Y_{sp}	set point value of the dependent variable

Greek

Δ increment
ε error signal
τ integral tuning constant

Superscripts

(n) n^{th} shot
(n + 1) $n + 1^{st}$ shot

Appendix III

Values of $y = \exp(-\eta)$

η	$e^{-\eta}$	η	$e^{-\eta}$	η	$e^{-\eta}$	η	$e^{-\eta}$	η	$e^{-\eta}$	η	$e^{-\eta}$
0.0	1.0000										
0.01	0.9900	0.41	0.6637	0.81	0.4449	1.21	0.2982	1.61	0.1999	2.01	0.1340
0.02	0.9802	0.42	0.6570	0.82	0.4404	1.22	0.2952	1.62	0.1979	2.02	0.1327
0.03	0.9704	0.43	0.6505	0.83	0.4306	1.23	0.2923	1.63	0.1959	2.03	0.1313
0.04	0.9608	0.44	0.6440	0.84	0.4317	1.24	0.2894	1.64	0.1940	2.04	0.1300
0.05	0.9512	0.45	0.6376	0.85	0.4274	1.25	0.2865	1.65	0.1920	2.05	0.1287
0.06	0.9418	0.46	0.6313	0.86	0.4232	1.26	0.2837	1.66	0.1901	2.06	0.1275
0.07	0.9324	0.47	0.6250	0.87	0.4190	1.27	0.2808	1.67	0.1882	2.07	0.1262
0.08	0.9231	0.48	0.6188	0.88	0.4148	1.28	0.2780	1.68	0.1864	2.08	0.1249
0.09	0.9139	0.49	0.6162	0.89	0.4107	1.29	0.2753	1.69	0.1845	2.09	0.1237
0.10	0.9048	0.50	0.6065	0.90	0.4066	1.30	0.2725	1.70	0.1827	2.10	0.1225
0.11	0.8958	0.51	0.6005	0.91	0.4025	1.31	0.2698	1.71	0.1809	2.11	0.1212
0.12	0.8869	0.52	0.5945	0.92	0.3985	1.32	0.2671	1.72	0.1791	2.12	0.1200
0.13	0.8781	0.53	0.5886	0.93	0.3946	1.33	0.2645	1.73	0.1773	2.13	0.1188
0.14	0.8694	0.54	0.5827	0.94	0.3906	1.34	0.2618	1.74	0.1775	2.14	0.1177
0.15	0.8607	0.55	0.5769	0.95	0.3867	1.35	0.2592	1.75	0.1738	2.15	0.1165
0.16	0.8521	0.56	0.5712	0.96	0.3829	1.36	0.2567	1.76	0.1720	2.16	0.1153
0.17	0.8437	0.57	0.5655	0.97	0.3791	1.37	0.2541	1.77	0.1703	2.17	0.1142
0.18	0.8353	0.58	0.5599	0.98	0.3753	1.38	0.2516	1.78	0.1686	2.18	0.1130
0.19	0.8270	0.59	0.5543	0.99	0.3716	1.39	0.2491	1.79	0.1670	2.19	0.1119
0.20	0.8187	0.60	0.5488	1.00	0.3679	1.40	0.2466	1.80	0.1653	2.20	0.1108
0.21	0.8106	0.61	0.5434	1.01	0.3642	1.41	0.2441	1.81	0.1637	2.21	0.1097
0.22	0.8025	0.62	0.5379	1.02	0.3606	1.42	0.2417	1.82	0.1620	2.22	0.1086
0.23	0.7945	0.63	0.5326	1.03	0.3570	1.43	0.2393	1.83	0.1604	2.23	0.1075
0.24	0.7866	0.64	0.5273	1.04	0.3535	1.44	0.2369	1.84	0.1588	2.24	0.1065
0.25	0.7788	0.65	0.5220	1.05	0.3499	1.45	0.2346	1.85	0.1572	2.25	0.1054
0.26	0.7711	0.66	0.5169	1.06	0.3465	1.46	0.2322	1.86	0.1557	2.26	0.1044
0.27	0.7634	0.67	0.5117	1.07	0.3430	1.47	0.2299	1.87	0.1541	2.27	0.1033
0.28	0.7558	0.68	0.5066	1.08	0.3396	1.48	0.2276	1.88	0.1526	2.28	0.1023
0.29	0.7483	0.69	0.5016	1.09	0.3362	1.49	0.2259	1.89	0.1511	2.29	0.1013
0.30	0.7408	0.70	0.4966	1.10	0.3329	1.50	0.2231	1.90	0.1496	2.30	0.1003

(*continues*)

η	$e^{-\eta}$	η	$e^{-\eta}$	η	$e^{-\eta}$	η	$e^{-\eta}$	η	$e^{-\eta}$	η	$e^{-\eta}$
0.31	0.7334	0.71	0.4916	1.11	0.3296	1.51	0.2209	1.91	0.1481	2.31	0.0993
0.32	0.7261	0.72	0.4868	1.12	0.3263	1.52	0.2187	1.92	0.1466	2.32	0.0983
0.33	0.7189	0.73	0.4819	1.13	0.3230	1.53	0.2165	1.93	0.1451	2.33	0.0973
0.34	0.7118	0.74	0.4771	1.14	0.3198	1.54	0.2144	1.94	0.1437	2.34	0.0963
0.35	0.7047	0.75	0.4724	1.15	0.3166	1.55	0.2122	1.95	0.1423	2.35	0.0954
0.36	0.6977	0.76	0.4677	1.16	0.3135	1.56	0.2101	1.96	0.1409	2.36	0.0944
0.37	0.6907	0.77	0.4630	1.17	0.3104	1.57	0.2080	1.97	0.1395	2.37	0.0935
0.38	0.6907	0.78	0.4584	1.18	0.3073	1.58	0.2060	1.98	0.1381	2.38	0.0926
0.39	0.6839	0.79	0.4538	1.19	0.3042	1.59	0.2039	1.99	0.1367	2.39	0.0916
0.40	0.6771	0.80	0.4493	1.20	0.3012	1.60	0.2019	2.00	0.1353	2.40	0.0907

Appendix IV

DPMO to Sigma to Yield % Conversion Table

DPMO	Sigma level	1.5 Sigma-shift yield %	DPMO	Sigma level	1.5 Sigma-shift yield %
3.40	6.008	99.99966	10,700.00	3.801	98.93000
5.00	5.914	99.99950	13,900.00	3.700	98.61000
5.42	5.896	99.99946	17,900.00	3.599	98.21000
8.00	5.821	99.99920	20,000.00	3.554	98.00000
8.55	5.802	99.99915	22,750.00	3.500	97.72500
10.00	5.765	99.99900	25,000.00	3.460	97.50000
13.35	5.700	99.99867	28,700.00	3.400	97.13000
15.00	5.672	99.99850	35,900.00	3.300	96.41000
20.00	5.607	99.99800	44,600.00	3.200	95.54000
30.00	5.514	99.99700	50,000.00	3.145	95.00000
31.70	5.500	99.99683	54,800.00	3.100	94.52000
40.00	5.444	99.99600	66,800.00	3.000	93.32000
48.10	5.400	99.99519	75,000.00	2.940	92.50000
50.00	5.391	99.99500	80,800.00	2.900	91.92000
60.00	5.346	99.99400	96,800.00	2.800	90.32000
70.00	5.309	99.99300	100,000.00	2.782	90.00000
72.40	5.300	99.99276	115,100.00	2.700	88.49000
80.00	5.274	99.99200	135,700.00	2.600	86.43000
90.00	5.246	99.99100	158,700.00	2.500	84.13000
100.00	5.219	99.99000	184,100.00	2.400	81.59000
108.00	5.200	99.98920	211,900.00	2.300	78.81000
150.00	5.116	99.98500	242,100.00	2.200	75.79000
159.00	5.101	99.98410	250,000.00	2.174	75.00000
200.00	5.040	99.98000	274,400.00	2.100	72.56000
232.00	5.001	99.97680	308,700.00	2.000	69.13000
233.00	4.999	99.97670	344,700.00	1.900	65.53000
300.00	4.932	99.97000	382,000.00	1.800	61.80000
340.00	4.898	99.96600	420,900.00	1.700	57.91000
400.00	4.853	99.96000	460,300.00	1.600	53.97000
485.00	4.799	99.95150	500,000.00	1.500	50.00000

(continues)

DPMO	Sigma level	1.5 Sigma-shift yield %	DPMO	Sigma level	1.5 Sigma-shift yield %
500.00	4.790	99.95000	540,000.00	1.400	46.00000
600.00	4.739	99.94000	579,200.00	1.300	42.08000
685.00	4.701	99.93150	618,000.00	1.200	38.20000
700.00	4.695	99.93000	655,600.00	1.100	34.44000
800.00	4.656	99.92000	691,500.00	1.000	30.85000
900.00	4.621	99.91000			
965.00	4.601	99.90350			
1,000.00	4.590	99.90000			
1,350.00	4.500	99.86500			
1,500.00	4.468	99.85000			
1,870.00	4.399	99.81300			
2,000.00	4.378	99.98000			
2,550.00	4.301	99.74500			
3,000.00	4.248	99.70000			
3,470.00	4.200	99.65300			
4,000.00	4.152	99.96000			
4,660.00	4.100	99.53400			
5,000.00	4.076	99.50000			
6,210.00	4.000	99.37900			
8,200.00	3.900	99.18000			
10,000.00	3.826	99.00000			

Appendix V

Standard Normal Distribution

The Z-Table contains the area under the standard normal curve from 0 to z. This can be used to compute the cumulative distribution values for the standard normal distribution.

The entries in the Z-Table are the probabilities that a standard normal random variable is between 0 and z (the indicated area).

Basically the area given in the Z-Table is $P(0 \leq Z \leq |a|)$, where a is a constant desired z-value (this is shown in the graph below for $a = 1.0$). This can be clarified by the simple example given in Chapter 3 (Example 3.1).

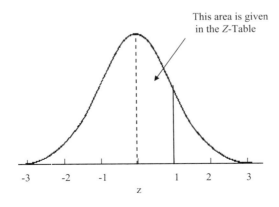

This area is given in the Z-Table

Z-Table

Critical Values for the Normal Distribution (Area under the Normal Curve from 0 to X)

Z	0.00000	0.01000	0.02000	0.03000	0.04000	0.05000	0.06000	0.07000	0.08000	0.09000
0.0	0.00000	0.00399	0.00798	0.01197	0.01595	0.01994	0.02392	0.02790	0.03188	0.03586
0.1	0.03983	0.04380	0.04776	0.05172	0.05567	0.05962	0.06356	0.06749	0.07142	0.07535
0.2	0.07926	0.08317	0.08706	0.09095	0.09483	0.09871	0.10257	0.10642	0.11026	0.11409
0.3	0.11791	0.12172	0.12552	0.12930	0.13307	0.13683	0.14058	0.14431	0.14803	0.15173
0.4	0.15542	0.15910	0.16276	0.16640	0.17003	0.17364	0.17724	0.18082	0.18439	0.18793
0.5	0.19146	0.19497	0.19847	0.20194	0.20540	0.20884	0.21226	0.21566	0.21904	0.22240
0.6	0.22575	0.22907	0.23237	0.23565	0.23891	0.24215	0.24537	0.24857	0.25175	0.25490
0.7	0.25804	0.26115	0.26424	0.26730	0.27035	0.27337	0.27637	0.27935	0.28230	0.28524
0.8	0.28814	0.29103	0.29389	0.29673	0.29955	0.30234	0.30511	0.30785	0.31057	0.31327
0.9	0.31594	0.31859	0.32121	0.32381	0.32639	0.32894	0.33147	0.33398	0.33646	0.33891
1.0	0.34134	0.34375	0.34614	0.34849	0.35083	0.35314	0.35543	0.35769	0.35993	0.36214
1.1	0.36433	0.36650	0.36864	0.37076	0.37286	0.37493	0.37698	0.37900	0.38100	0.38298
1.2	0.38493	0.38686	0.38877	0.39065	0.39251	0.39435	0.39617	0.39796	0.39973	0.40147
1.3	0.40320	0.40490	0.40658	0.40824	0.40988	0.41149	0.41308	0.41466	0.41621	0.41774
1.4	0.41924	0.42073	0.42220	0.42364	0.42507	0.42647	0.42785	0.42922	0.43056	0.43189
1.5	0.43319	0.43448	0.43574	0.43699	0.43822	0.43943	0.44062	0.44179	0.44295	0.44408
1.6	0.44520	0.44630	0.44738	0.44845	0.44950	0.45053	0.45154	0.45254	0.45352	0.45449
1.7	0.45543	0.45637	0.45728	0.45818	0.45907	0.45994	0.46080	0.46164	0.46246	0.46327
1.8	0.46407	0.46485	0.46562	0.46638	0.46712	0.46784	0.46856	0.46926	0.46995	0.47062
1.9	0.47128	0.47193	0.47257	0.47320	0.47381	0.47441	0.47500	0.47558	0.47615	0.47670

z	0.00	0.01	0.02	0.03	0.04	0.05	0.06	0.07	0.08	0.09
2.0	0.47725	0.47778	0.47831	0.47882	0.47932	0.47982	0.48030	0.48077	0.48124	0.48169
2.1	0.48214	0.48257	0.48300	0.48341	0.48382	0.48422	0.48461	0.48500	0.48537	0.48574
2.2	0.48610	0.48645	0.48679	0.48713	0.48745	0.48778	0.48809	0.48840	0.48870	0.48899
2.3	0.48928	0.48956	0.48983	0.49010	0.49036	0.49061	0.49086	0.49111	0.49134	0.49158
2.4	0.49180	0.49202	0.49224	0.49245	0.49266	0.49286	0.49305	0.49324	0.49343	0.49361
2.5	0.49379	0.49396	0.49413	0.49430	0.49446	0.49461	0.49477	0.49492	0.49506	0.49520
2.6	0.49534	0.49547	0.49560	0.49573	0.49585	0.49598	0.49609	0.49621	0.49632	0.49643
2.7	0.49653	0.49664	0.49674	0.49683	0.49693	0.49702	0.49711	0.49720	0.49728	0.49736
2.8	0.49744	0.49752	0.49760	0.49767	0.49774	0.49781	0.49788	0.49795	0.49801	0.49807
2.9	0.49813	0.49819	0.49825	0.49831	0.49836	0.49841	0.49846	0.49851	0.49856	0.49861
3.0	0.49865	0.49869	0.49874	0.49882	0.49882	0.49886	0.49889	0.49893	0.49896	0.49900
3.1	0.49903	0.49906	0.49910	0.49913	0.49916	0.49918	0.49921	0.49924	0.49926	0.49929
3.2	0.49931	0.49934	0.49936	0.49938	0.49940	0.49942	0.49944	0.49946	0.49948	0.49950
3.3	0.49952	0.49953	0.49955	0.49957	0.49958	0.49960	0.49961	0.49962	0.49964	0.49965
3.4	0.49966	0.49968	0.49969	0.49970	0.49971	0.49972	0.49973	0.49974	0.49975	0.49976
3.5	0.49977	0.49978	0.49978	0.49979	0.49980	0.49981	0.49981	0.49982	0.49983	0.49983
3.6	0.49984	0.49985	0.49985	0.49986	0.49986	0.49987	0.49987	0.49988	0.49988	0.49989
3.7	0.49989	0.49990	0.49990	0.49990	0.49991	0.49991	0.49992	0.49992	0.49992	0.49992
3.8	0.49993	0.49993	0.49993	0.49994	0.49994	0.49994	0.49994	0.49995	0.49995	0.49995
3.9	0.49995	0.49995	0.49996	0.49996	0.49996	0.49996	0.49996	0.49996	0.49997	0.49997
4.0	0.49997	0.49997	0.49997	0.49997	0.49997	0.49997	0.49998	0.49998	0.49998	0.49998

Adapted from the National Institute of Standards and Technology Engineering Statistics Handbook.

Appendix VI

Critical Values of *t*-Distribution

This table contains the upper critical values of the student's *t*-distribution. The upper critical values are computed using the percent point function. Due to the symmetry of the *t*-distribution, this table can be used for both one-sided (lower and upper) and two-sided tests using the appropriate value of α.

The significance level α is demonstrated with the following graph that plots a *t*-distribution with 10 degrees of freedom. The most commonly used significance level is $\alpha = 0.05$. For a two-sided test, we compute the percent point function at $\alpha/2$ (0.025). If the absolute value of the test statistic is greater than the upper critical value (0.025), then we reject the null hypothesis. Due to the symmetry of the *t*-distribution, we only tabulate the upper critical values in the table Appendix VI.

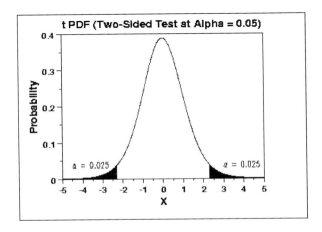

Given a specified value α

1. For a two-sided test, find the column corresponding to $\alpha/2$ and reject the null hypothesis if the absolute value of the test statistic is greater than the value of $t_{\alpha/2}$ in the following table.
2. For an upper one-sided test, find the column corresponding to α and reject the null hypothesis if the test statistic is greater than the tabled value.
3. For a lower one-sided test, find the column corresponding to α and reject the null hypothesis if the test statistic is less than the negative of the tabled value.

Upper Critical Values of Student's t-Distribution with *df* Degrees of Freedom
Probability of Exceeding the Critical Value

df	$t_{0.10}$	$t_{0.05}$	$t_{0.025}$	$t_{0.01}$	$t_{0.005}$	$t_{0.001}$
1	3.078	6.314	12.706	31.821	63.657	318.313
2	1.886	2.920	4.303	6.965	9.925	22.327
3	1.638	2.353	3.182	4.541	5.841	10.215
4	1.533	2.132	2.776	3.747	4.604	7.173
5	1.476	2.015	2.571	3.365	4.032	5.893
6	1.440	1.943	2.447	3.143	3.707	5.208
7	1.415	1.895	2.365	2.998	3.499	4.782
8	1.397	1.860	2.306	2.896	3.355	4.499
9	1.383	1.833	2.262	2.821	3.250	4.296
10	1.372	1.812	2.228	2.764	3.169	4.143
11	1.363	1.796	2.201	2.718	3.106	4.024
12	1.356	1.782	2.179	2.681	3.055	3.929
13	1.350	1.771	2.160	2.650	3.012	3.852
14	1.345	1.761	2.145	2.624	2.977	3.787
15	1.341	1.753	2.131	2.602	2.947	3.733
16	1.337	1.746	2.120	2.583	2.921	3.686
17	1.333	1.740	2.110	2.567	2.898	3.646
18	1.330	1.734	2.101	2.552	2.878	3.610
19	1.328	1.729	2.093	2.539	2.861	3.579
20	1.325	1.725	2.086	2.528	2.845	3.552
21	1.323	1.721	2.080	2.518	2.831	3.527
22	1.321	1.717	2.074	2.508	2.819	3.505
23	1.319	1.714	2.069	2.500	2.807	3.485
24	1.318	1.711	2.064	2.492	2.797	3.467
25	1.316	1.708	2.060	2.485	2.787	3.450
26	1.315	1.706	2.056	2.479	2.779	3.435
27	1.314	1.703	2.052	2.473	2.771	3.421

(continues)

Probability of Exceeding the Critical Value (*continued*)

df	$t_{0.10}$	$t_{0.05}$	$t_{0.025}$	$t_{0.01}$	$t_{0.005}$	$t_{0.001}$
28	1.313	1.701	2.048	2.467	2.763	3.408
29	1.311	1.699	2.045	2.462	2.756	3.396
30	1.310	1.697	2.042	2.457	2.750	3.385
31	1.309	1.696	2.040	2.453	2.744	3.375
32	1.309	1.694	2.037	2.449	2.738	3.365
33	1.308	1.692	2.035	2.445	2.733	3.356
34	1.307	1.691	2.032	2.441	2.728	3.348
35	1.306	1.690	2.030	2.438	2.724	3.340
36	1.306	1.688	2.028	2.434	2.719	3.333
37	1.305	1.687	2.026	2.431	2.715	3.326
38	1.304	1.686	2.024	2.429	2.712	3.319
39	1.304	1.685	2.023	2.426	2.708	3.313
40	1.303	1.684	2.021	2.423	2.704	3.307
41	1.303	1.683	2.020	2.421	2.701	3.301
42	1.302	1.682	2.018	2.418	2.698	3.296
43	1.302	1.681	2.017	2.416	2.695	3.291
44	1.301	1.680	2.015	2.414	2.692	3.286
45	1.301	1.679	2.014	2.412	2.690	3.281
46	1.300	1.679	2.013	2.410	2.687	3.277
47	1.300	1.678	2.012	2.408	2.685	3.273
48	1.299	1.677	2.011	2.407	2.682	3.269
49	1.299	1.677	2.010	2.405	2.680	3.265
50	1.299	1.676	2.009	2.403	2.678	3.261
51	1.298	1.675	2.008	2.402	2.676	3.258
52	1.298	1.675	2.007	2.400	2.674	3.255
53	1.298	1.674	2.006	2.399	2.672	3.251
54	1.297	1.674	2.005	2.397	2.670	3.248
55	1.297	1.673	2.004	2.396	2.668	3.245
56	1.297	1.673	2.003	2.395	2.667	3.242
57	1.297	1.672	2.002	2.394	2.665	3.239
58	1.296	1.672	2.002	2.392	2.663	3.237
59	1.296	1.671	2.001	2.391	2.662	3.234
60	1.296	1.671	2.000	2.390	2.660	3.232
61	1.296	1.670	2.000	2.389	2.659	3.229
62	1.295	1.670	1.999	2.388	2.657	3.227
63	1.295	1.669	1.998	2.387	2.656	3.225
64	1.295	1.669	1.998	2.386	2.655	3.223
65	1.295	1.669	1.997	2.385	2.654	3.220
66	1.295	1.668	1.997	2.384	2.652	3.218
67	1.294	1.668	1.996	2.383	2.651	3.216
68	1.294	1.668	1.995	2.382	2.650	3.214
69	1.294	1.667	1.995	2.382	2.649	3.213
70	1.294	1.667	1.994	2.381	2.648	3.211
71	1.294	1.667	1.994	2.380	2.647	3.209
72	1.293	1.666	1.993	2.379	2.646	3.207

(*continues*)

Probability of Exceeding the Critical Value (*continued*)

df	$t_{0.10}$	$t_{0.05}$	$t_{0.025}$	$t_{0.01}$	$t_{0.005}$	$t_{0.001}$
73	1.293	1.666	1.993	2.379	2.645	3.206
74	1.293	1.666	1.993	2.378	2.644	3.204
75	1.293	1.665	1.992	2.377	2.643	3.202
76	1.293	1.665	1.992	2.376	2.642	3.201
77	1.293	1.665	1.991	2.376	2.641	3.199
78	1.292	1.665	1.991	2.375	2.640	3.198
79	1.292	1.664	1.990	2.374	2.640	3.197
80	1.292	1.664	1.990	2.374	2.639	3.195
81	1.292	1.664	1.990	2.373	2.638	3.194
82	1.292	1.664	1.989	2.373	2.637	3.193
83	1.292	1.663	1.989	2.372	2.636	3.191
84	1.292	1.663	1.989	2.372	2.636	3.190
85	1.292	1.663	1.988	2.371	2.635	3.189
86	1.291	1.663	1.988	2.370	2.634	3.188
87	1.291	1.663	1.988	2.370	2.634	3.187
88	1.291	1.662	1.987	2.369	2.633	3.185
89	1.291	1.662	1.987	2.369	2.632	3.184
90	1.291	1.662	1.987	2.368	2.632	3.183
91	1.291	1.662	1.986	2.368	2.631	3.182
92	1.291	1.662	1.986	2.368	2.630	3.181
93	1.291	1.661	1.986	2.367	2.630	3.180
94	1.291	1.661	1.986	2.367	2.629	3.179
95	1.291	1.661	1.985	2.366	2.629	3.178
96	1.290	1.661	1.985	2.366	2.628	3.177
97	1.290	1.661	1.985	2.365	2.627	3.176
98	1.290	1.661	1.984	2.365	2.627	3.175
99	1.290	1.660	1.984	2.365	2.626	3.175
100	1.290	1.660	1.984	2.364	2.626	3.174
∞	1.282	1.645	1.960	2.326	2.576	3.090

Adapted from the National Institute of Standards and Technology Engineering Statistics Handbook.

Appendix VII

Critical Values of Chi-Square Distribution
with Degrees of Freedom

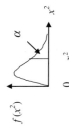

$$f(x^2)$$

$$\alpha$$

$$0 \qquad x_\alpha^2 \qquad x^2$$

Chi-Square $\chi^2_{\alpha, df}$

							α				
df	0.9950	0.9900	0.9750	0.9500	0.9000	0.5000	0.1000	0.0500	0.0250	0.0100	0.0050
1	0.0000	0.0001	0.0009	0.0039	0.0157	0.4549	2.7055	3.8414	5.0238	6.6349	7.8794
2	0.0100	0.0201	0.0506	0.1025	0.2107	1.3862	4.6051	5.9914	7.3777	9.2103	10.596
3	0.0717	0.1148	0.2157	0.3518	0.5843	2.3659	6.2513	7.8147	9.3484	11.344	12.838
4	0.2069	0.2971	0.4844	0.7107	1.0636	3.3566	7.7794	9.4877	11.143	13.276	14.860
5	0.4117	0.5542	0.8312	1.1454	1.6103	4.3514	9.2363	11.070	12.832	15.086	16.749
6	0.6757	0.8720	1.2373	1.6353	2.2041	5.3481	10.646	12.591	14.449	16.811	18.547
7	0.9892	1.2390	1.6898	2.1673	2.8331	6.3458	12.017	14.067	16.012	18.475	20.277
8	1.3444	1.6464	2.1797	2.7326	3.4895	7.3441	13.361	15.507	17.534	20.090	21.955
9	1.7349	2.0879	2.7003	3.3251	4.1681	8.3428	14.683	16.919	19.022	21.666	23.589
10	2.1558	2.5582	3.2469	3.9403	4.8651	9.3418	15.987	18.307	20.483	23.209	25.188
11	2.6032	3.0534	3.8157	4.5748	5.5777	10.341	17.275	19.675	21.920	24.725	26.756
12	3.0738	3.5705	4.4037	5.2260	6.3038	11.340	18.549	21.026	23.336	26.217	28.299
13	3.5650	4.1069	5.0087	5.8918	7.0415	12.340	19.811	22.362	24.735	27.688	29.819
14	4.0746	4.6604	5.6287	6.5706	7.7895	13.339	21.064	23.684	26.119	29.141	31.319
15	4.6009	5.2293	6.2621	7.2609	8.5467	14.338	22.307	24.995	27.488	30.577	32.801
16	5.1422	5.8122	6.9076	7.9616	9.3122	15.338	23.541	26.296	28.845	31.999	34.267
17	5.6972	6.4077	7.5641	8.6717	10.085	16.338	24.769	27.587	30.191	33.408	35.718
18	6.2648	7.0149	8.2307	9.3904	10.864	17.338	25.989	28.869	31.526	34.805	37.156
19	6.8439	7.6327	8.9065	10.117	11.650	18.337	27.203	30.143	32.852	36.190	38.582
20	7.4338	8.2604	9.5908	10.850	12.442	19.337	28.412	31.410	34.169	37.556	39.996
21	8.0336	8.8972	10.282	11.591	13.239	20.337	29.615	32.670	35.478	38.932	41.401
22	8.6427	9.5424	10.982	12.338	14.041	21.337	30.813	33.924	36.780	40.289	42.795
23	9.2604	10.195	11.688	13.090	14.847	22.337	32.006	35.172	38.075	41.638	44.181
24	9.8862	10.856	12.401	13.848	15.653	23.337	33.196	36.415	39.364	42.979	45.558
25	10.519	11.524	13.119	0.9500	16.473	24.336	34.381	37.625	40.646	44.314	46.927
26	11.160	12.198	13.843	15.379	17.291	25.336	35.563	38.885	41.923	45.641	48.289

27	11.807	12.878	14.573	16.151	18.113	26.336	36.741	40.113	43.194	46.963	49.644
28	12.461	13.564	15.307	16.927	18.939	27.336	37.915	41.337	44.460	48.278	50.993
29	13.121	14.256	16.047	17.708	19.767	28.336	39.087	42.556	45.722	49.587	52.335
30	13.786	14.953	16.790	18.492	20.599	29.336	40.256	43.772	46.979	50.892	53.672
31	14.458	15.655	17.539	19.281	21.434	30.336	41.422	44.985	48.232	52.191	55.003
32	15.134	16.362	18.291	20.072	22.271	31.336	42.585	46.194	49.480	53.486	56.328
33	15.815	17.074	19.047	20.867	23.110	32.336	43.745	47.400	50.725	54.776	57.648
34	16.501	17.789	19.806	21.664	23.952	33.335	44.903	48.602	51.966	56.061	58.964
35	17.192	18.509	20.569	22.465	24.797	34.335	46.059	49.802	53.203	57.342	60.275
36	17.887	19.336	21.336	23.269	25.643	35.335	47.212	50.998	54.437	58.619	61.581
37	18.586	19.960	22.106	24.075	26.492	36.335	48.363	52.192	55.668	59.893	62.883
38	19.289	20.691	22.878	24.884	27.343	37.335	49.513	53.384	56.896	61.162	64.181
39	19.996	21.426	23.654	25.695	28.196	38.335	50.660	54.572	58.120	62.428	65.475
40	20.706	22.164	24.433	26.509	29.050	39.335	51.805	55.758	59.341	63.691	66.766
41	21.421	22.906	25.215	27.326	29.907	40.335	52.949	56.942	60.561	64.950	68.053
42	22.138	23.650	25.999	28.144	30.765	41.335	54.090	58.124	61.777	66.206	69.336
43	22.859	24.398	26.785	28.965	31.625	42.335	55.230	59.304	62.990	67.459	70.616
44	23.587	25.148	27.575	29.787	32.487	43.335	56.369	60.481	64.201	68.710	71.892
45	24.311	25.901	28.366	30.612	33.350	44.335	57.505	61.656	65.410	69.957	73.166
46	25.041	26.657	29.160	31.439	34.215	45.335	58.641	62.830	66.617	71.201	74.436
47	25.775	27.416	29.956	32.268	35.081	46.335	59.774	64.001	67.821	72.443	75.704
48	26.510	28.177	30.755	33.098	35.949	47.335	60.907	65.171	69.023	73.683	76.969
49	27.249	28.914	31.555	33.930	36.818	48.334	62.038	66.339	70.222	74.919	78.231
50	27.991	29.582	32.357	34.764	37.689	49.334	63.167	67.505	71.420	76.154	79.490
51	28.735	30.475	33.162	35.600	38.560	50.335	64.295	68.669	72.616	77.386	80.747
52	29.481	31.246	33.968	36.437	39.433	51.335	65.422	69.832	73.810	78.616	82.001
53	30.230	32.018	34.776	37.276	40.308	52.335	66.548	70.993	75.002	79.843	83.252

(continues)

Chi-Square $\chi^2_{\alpha,df}$ (continued)

df \ α	0.9950	0.9900	0.9750	0.9500	0.9000	0.5000	0.1000	0.0500	0.0250	0.0100	0.0050
54	30.981	32.793	35.586	38.116	41.183	53.335	67.673	72.153	76.192	81.069	84.502
55	31.735	33.570	36.398	38.958	42.060	54.335	68.796	73.311	77.380	82.292	85.749
56	32.490	34.350	37.212	39.801	42.937	55.335	69.919	74.468	78.567	83.513	86.994
57	33.248	35.027	38.027	40.646	43.816	56.335	71.040	75.624	79.756	84.733	88.236
58	34.008	38.844	38.844	41.492	44.696	57.335	72.160	76.778	80.936	85.950	89.477
59	34.770	36.698	39.662	42.339	45.577	58.335	73.279	77.931	82.117	87.166	90.715
60	35.534	37.485	40.482	43.188	46.459	59.335	74.397	79.082	83.298	88.379	91.952
61	36.300	38.273	41.303	44.038	47.342	60.335	75.514	80.232	84.476	89.591	93.186
62	37.068	39.063	42.126	44.889	48.226	61.335	76.630	81.381	85.654	90.802	94.419
63	37.838	39.855	42.950	45.741	49.111	62.335	77.745	82.529	86.830	92.010	95.649
64	38.610	40.649	43.776	46.595	49.996	63.334	78.860	83.675	88.004	93.217	96.878
65	39.383	41.444	44.603	47.450	50.883	64.334	79.973	84.821	89.177	94.422	98.105
66	40.158	42.240	45.431	48.305	51.770	65.334	81.085	85.965	90.349	95.626	99.330
67	40.935	43.038	46.261	49.162	52.659	66.334	82.197	87.108	91.519	96.828	100.55
68	41.713	43.838	47.092	50.020	53.548	67.334	83.308	88.250	92.689	98.028	101.78
69	42.493	44.639	47.924	50.879	54.438	68.334	84.418	89.391	93.856	99.228	102.99
70	43.275	45.442	48.758	51.739	55.329	69.334	85.527	90.531	95.023	100.42	104.21
71	44.058	46.246	49.592	52.600	56.221	70.334	86.635	91.670	96.189	101.62	105.43
72	44.843	47.051	50.428	53.462	57.113	71.334	87.743	92.808	97.353	102.81	106.65
73	45.629	47.858	51.265	54.325	58.006	72.334	88.850	93.945	98.516	104.01	107.86
74	46.417	48.666	52.103	55.189	58.900	73.334	89.956	95.081	99.678	105.20	109.07
75	47.206	49.475	52.942	56.054	59.795	74.334	91.061	96.217	100.84	106.39	110.28
76	47.996	50.286	53.782	56.920	60.690	75.334	92.166	97.351	101.99	107.58	111.49
77	48.788	51.097	54.623	57.786	61.586	76.334	93.270	98.484	103.16	108.77	112.70
78	49.581	51.910	55.466	58.654	62.483	77.334	94.374	99.617	104.31	109.96	113.91

79	50.376	52.725	56.309	59.522	63.380	78.334	95.476	100.75	105.47	111.14	115.12
80	51.172	53.540	57.153	60.391	64.278	79.334	96.578	101.88	106.63	112.33	116.32
81	51.969	54.357	57.998	61.261	65.176	80.334	97.680	103.01	107.78	113.51	117.52
82	52.767	55.174	58.845	62.132	66.076	81.334	98.780	104.14	108.94	114.69	118.73
83	53.567	55.993	59.692	63.004	66.976	82.334	99.880	105.27	110.09	115.87	119.93
84	54.367	56.813	60.540	63.876	67.876	83.334	100.98	106.40	111.24	117.06	121.13
85	55.170	57.634	61.389	64.749	68.777	84.334	102.08	107.52	112.39	118.23	122.32
86	55.973	58.456	62.239	65.623	69.679	85.334	103.18	108.65	113.54	119.41	123.52
87	56.777	59.279	63.089	66.498	70.581	86.334	104.28	109.77	114.69	120.59	124.72
88	57.582	60.103	63.941	67.373	71.484	87.334	105.37	110.90	115.84	121.76	125.91
89	58.389	60.928	64.793	68.249	72.387	88.334	106.47	112.02	116.99	122.94	127.11
90	59.196	61.754	65.647	69.126	73.291	89.334	107.57	113.14	118.14	124.11	128.30
91	60.005	62.581	66.501	70.003	74.196	90.334	108.66	114.27	119.28	125.29	129.49
92	60.814	63.409	67.356	70.882	75.100	91.334	109.76	115.39	120.43	126.46	130.68
93	61.625	64.238	68.211	71.760	76.006	92.334	110.85	116.51	121.57	127.63	131.87
94	62.437	65.068	69.068	72.640	76.912	93.334	111.94	117.63	122.71	128.80	133.06
95	63.250	65.898	69.925	73.520	77.818	94.334	113.04	118.75	123.86	129.97	134.25
96	64.063	66.730	70.783	74.401	78.725	95.334	114.13	119.87	125.00	131.14	135.43
97	64.878	67.562	71.642	75.282	79.633	96.334	115.22	120.99	126.14	132.31	136.62
98	65.693	68.396	72.501	76.164	80.541	97.334	116.31	122.11	127.28	133.48	137.80
99	66.510	69.230	73.361	77.046	81.449	98.334	117.41	123.22	128.42	134.64	138.98
100	67.327	70.065	74.222	77.929	82.358	99.334	118.50	124.34	129.56	135.81	140.17

Appendix VIII

Upper Critical Values of the F-Distribution for df_1 Numerator Degrees of Freedom and df_2 Denominator Degrees of Freedom

How to Use the F Table

The F-Table contains the upper critical values of the F-distribution. This table is used for one-sided F-tests at the $\alpha = 0.05$, 0.10, and 0.01 levels.

More specifically, a test statistic is computed with df_1 (df error) and df_2 (df factor) degrees of freedom, and the result is compared to the table in Appendix VII. For a one-sided test, the null hypothesis is rejected when the test statistic is greater than the tabled value. This is demonstrated with the graph of an F-distribution with $df_1 = 10$ and $df_2 = 10$. The shaded area of the graph indicates the rejection region at the α significance level. Since this is a one-sided test, we have α probability in the upper tail of exceeding the critical value and zero in the lower tail. Because the F-distribution is asymmetric, a two-sided test requires a set of tables (not included here) that contain the rejection regions for both the lower and upper tails.

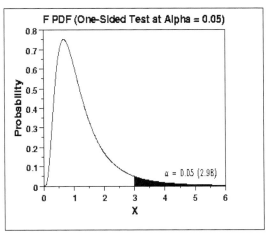

Appendix VIII

Upper Critical Values of the F-Distribution for df_1 Numerator Degrees of Freedom and df_2 Denominator Degrees of Freedom

10% Significance Level

$$F_{0.10}\ (df_1, df_2)$$

df_2 \ df_1	1	2	3	4	5	6	7	8	9	10
1	39.863	49.500	53.593	55.833	57.240	58.204	58.906	59.439	59.858	60.195
2	8.526	9.000	9.162	9.243	9.293	9.326	9.349	9.367	9.381	9.392
3	5.538	5.462	5.391	5.343	5.309	5.285	5.266	5.252	5.240	5.230
4	4.545	4.325	4.191	4.107	4.051	4.010	3.979	3.955	3.936	3.920
5	4.060	3.780	3.619	3.520	3.453	3.405	3.368	3.339	3.316	3.297
6	3.776	3.463	3.289	3.181	3.108	3.055	3.014	2.983	2.958	2.937
7	3.589	3.257	3.074	2.961	2.883	2.827	2.785	2.752	2.725	2.703
8	3.458	3.113	2.924	2.806	2.726	2.668	2.624	2.589	2.561	2.538
9	3.360	3.006	2.813	2.693	2.611	2.551	2.505	2.469	2.440	2.416
10	3.285	2.924	2.728	2.605	2.522	2.461	2.414	2.377	2.347	2.323
11	3.225	2.860	2.660	2.536	2.451	2.389	2.342	2.304	2.274	2.248
12	3.177	2.807	2.606	2.480	2.394	2.331	2.283	2.245	2.214	2.188
13	3.136	2.763	2.560	2.434	2.347	2.283	2.234	2.195	2.164	2.138
14	3.102	2.726	2.522	2.395	2.307	2.243	2.193	2.154	2.122	2.095
15	3.073	2.695	2.490	2.361	2.273	2.208	2.158	2.119	2.086	2.059
16	3.048	2.668	2.462	2.333	2.244	2.178	2.128	2.088	2.055	2.028
17	3.026	2.645	2.437	2.308	2.218	2.152	2.102	2.061	2.028	2.001
18	3.007	2.624	2.416	2.286	2.196	2.130	2.079	2.038	2.005	1.977
19	2.990	2.606	2.397	2.266	2.176	2.109	2.058	2.017	1.984	1.956
20	2.975	2.589	2.380	2.249	2.158	2.091	2.040	1.999	1.965	1.937
21	2.961	2.575	2.365	2.233	2.142	2.075	2.023	1.982	1.948	1.920
22	2.949	2.561	2.351	2.219	2.128	2.060	2.008	1.967	1.933	1.904
23	2.937	2.549	2.339	2.207	2.115	2.047	1.995	1.953	1.919	1.890
24	2.927	2.538	2.327	2.195	2.103	2.035	1.983	1.941	1.906	1.877
25	2.918	2.528	2.317	2.184	2.092	2.024	1.971	1.929	1.895	1.866
26	2.909	2.519	2.307	2.174	2.082	2.014	1.961	1.919	1.884	1.855

27	2.901	2.511	2.299	2.165	2.073	2.005	1.952	1.909	1.874	1.845
28	2.894	2.503	2.291	2.157	2.064	1.996	1.943	1.900	1.865	1.836
29	2.887	2.495	2.283	2.149	2.057	1.988	1.935	1.892	1.857	1.827
30	2.881	2.489	2.276	2.142	2.049	1.980	1.927	1.884	1.849	1.819
31	2.875	2.482	2.270	2.136	2.042	1.973	1.920	1.877	1.842	1.812
32	2.869	2.477	2.263	2.129	2.036	1.967	1.913	1.870	1.835	1.805
33	2.864	2.471	2.258	2.123	2.030	1.961	1.907	1.864	1.828	1.799
34	2.859	2.466	2.252	2.118	2.024	1.955	1.901	1.858	1.822	1.793
35	2.855	2.461	2.247	2.113	2.019	1.950	1.896	1.852	1.817	1.787
36	2.850	2.456	2.243	2.108	2.014	1.945	1.891	1.847	1.811	1.781
37	2.846	2.452	2.238	2.103	2.009	1.940	1.886	1.842	1.806	1.776
38	2.842	2.448	2.234	2.099	2.005	1.935	1.881	1.838	1.802	1.772
39	2.839	2.444	2.230	2.095	2.001	1.931	1.877	1.833	1.797	1.767
40	2.835	2.440	2.226	2.091	1.997	1.927	1.873	1.829	1.793	1.763
41	2.832	2.437	2.222	2.087	1.993	1.923	1.869	1.825	1.789	1.759
42	2.829	2.434	2.219	2.084	1.989	1.919	1.865	1.821	1.785	1.755
43	2.826	2.430	2.216	2.080	1.986	1.916	1.861	1.817	1.781	1.751
44	2.823	2.427	2.213	2.077	1.983	1.913	1.858	1.814	1.778	1.747
45	2.820	2.425	2.210	2.074	1.980	1.909	1.855	1.811	1.774	1.744
46	2.818	2.422	2.207	2.071	1.977	1.906	1.852	1.808	1.771	1.741
47	2.815	2.419	2.204	2.068	1.974	1.903	1.849	1.805	1.768	1.738
48	2.813	2.417	2.202	2.066	1.971	1.901	1.846	1.802	1.765	1.735
49	2.811	2.414	2.199	2.063	1.968	1.898	1.843	1.799	1.763	1.732
50	2.809	2.412	2.197	2.061	1.966	1.895	1.840	1.796	1.760	1.729
60	2.791	2.393	2.177	2.041	1.946	1.875	1.819	1.775	1.738	1.707
70	2.779	2.380	2.164	2.027	1.931	1.860	1.804	1.760	1.723	1.691
80	2.769	2.370	2.154	2.016	1.921	1.849	1.793	1.748	1.711	1.680
90	2.762	2.363	2.146	2.008	1.912	1.841	1.785	1.739	1.702	1.670
100	2.756	2.356	2.139	2.002	1.906	1.834	1.778	1.732	1.695	1.663

Appendix VIII (continued)

Upper Critical Values of the F-Distribution for df_1 Numerator Degrees of Freedom and df_2 Denominator Degrees of Freedom

10% Significance Level

$$F_{0.10}\,(df_1, df_2)$$

df_2 \ df_1	11	12	13	14	15	16	17	18	19	20
1	60.473	60.705	60.903	61.073	61.220	61.350	61.464	61.566	61.658	61.740
2	9.401	9.408	9.415	9.420	9.425	9.429	9.433	9.436	9.439	9.441
3	5.222	5.216	5.210	5.205	5.200	5.196	5.193	5.190	5.187	5.184
4	3.907	3.896	3.886	3.878	3.870	3.864	3.858	3.853	3.849	3.844
5	3.282	3.268	3.257	3.247	3.238	3.230	3.223	3.217	3.212	3.207
6	2.920	2.905	2.892	2.881	2.871	2.863	2.855	2.848	2.842	2.836
7	2.684	2.668	2.654	2.643	2.632	2.623	2.615	2.607	2.601	2.595
8	2.519	2.502	2.488	2.475	2.464	2.455	2.446	2.438	2.431	2.425
9	2.396	2.379	2.364	2.351	2.340	2.329	2.320	2.312	2.305	2.298
10	2.302	2.284	2.269	2.255	2.244	2.233	2.224	2.215	2.208	2.201
11	2.227	2.209	2.193	2.179	2.167	2.156	2.147	2.138	2.130	2.123
12	2.166	2.147	2.131	2.117	2.105	2.094	2.084	2.075	2.067	2.060
13	2.116	2.097	2.080	2.066	2.053	2.042	2.032	2.023	2.014	2.007
14	2.073	2.054	2.037	2.022	2.010	1.998	1.988	1.978	1.970	1.962
15	2.037	2.017	2.000	1.985	1.972	1.961	1.950	1.941	1.932	1.924
16	2.005	1.985	1.968	1.953	1.940	1.928	1.917	1.908	1.899	1.891
17	1.978	1.958	1.940	1.925	1.912	1.900	1.889	1.879	1.870	1.862
18	1.954	1.933	1.916	1.900	1.887	1.875	1.864	1.854	1.845	1.837
19	1.932	1.912	1.894	1.878	1.865	1.852	1.841	1.831	1.822	1.814
20	1.913	1.892	1.875	1.859	1.845	1.833	1.821	1.811	1.802	1.794
21	1.896	1.875	1.857	1.841	1.827	1.815	1.803	1.793	1.784	1.776
22	1.880	1.859	1.841	1.825	1.811	1.798	1.787	1.777	1.768	1.759
23	1.866	1.845	1.827	1.811	1.796	1.784	1.772	1.762	1.753	1.744
24	1.853	1.832	1.814	1.797	1.783	1.770	1.759	1.748	1.739	1.730
25	1.841	1.820	1.802	1.785	1.771	1.758	1.746	1.736	1.726	1.718
26	1.830	1.809	1.790	1.774	1.760	1.747	1.735	1.724	1.715	1.706

27	1.695	1.704	1.714	1.724	1.736	1.749	1.764	1.780	1.799	1.820
28	1.685	1.694	1.704	1.715	1.726	1.740	1.754	1.771	1.790	1.811
29	1.676	1.685	1.695	1.705	1.717	1.731	1.745	1.762	1.781	1.802
30	1.667	1.676	1.686	1.697	1.709	1.722	1.737	1.754	1.773	1.794
31	1.659	1.668	1.678	1.689	1.701	1.714	1.729	1.746	1.765	1.787
32	1.652	1.661	1.671	1.682	1.694	1.707	1.722	1.739	1.758	1.780
33	1.645	1.654	1.664	1.675	1.687	1.700	1.715	1.732	1.751	1.773
34	1.638	1.647	1.657	1.668	1.680	1.694	1.709	1.726	1.745	1.767
35	1.632	1.641	1.651	1.662	1.674	1.688	1.703	1.720	1.739	1.761
36	1.626	1.635	1.645	1.656	1.669	1.682	1.697	1.715	1.734	1.756
37	1.620	1.630	1.640	1.651	1.663	1.677	1.692	1.709	1.729	1.751
38	1.615	1.624	1.635	1.646	1.658	1.672	1.687	1.704	1.724	1.746
39	1.610	1.619	1.630	1.641	1.653	1.667	1.682	1.700	1.719	1.741
40	1.605	1.615	1.625	1.636	1.649	1.662	1.678	1.695	1.715	1.737
41	1.601	1.610	1.620	1.632	1.644	1.658	1.673	1.691	1.710	1.733
42	1.596	1.606	1.616	1.628	1.640	1.654	1.669	1.687	1.706	1.729
43	1.592	1.602	1.612	1.624	1.636	1.650	1.665	1.683	1.703	1.725
44	1.588	1.598	1.608	1.620	1.632	1.646	1.662	1.679	1.699	1.721
45	1.585	1.594	1.605	1.616	1.629	1.643	1.658	1.676	1.695	1.718
46	1.581	1.591	1.601	1.613	1.625	1.639	1.655	1.672	1.692	1.715
47	1.578	1.587	1.598	1.609	1.622	1.636	1.652	1.669	1.689	1.712
48	1.574	1.584	1.594	1.606	1.619	1.633	1.648	1.666	1.686	1.709
49	1.571	1.581	1.591	1.603	1.616	1.630	1.645	1.663	1.683	1.706
50	1.568	1.578	1.588	1.600	1.613	1.627	1.643	1.660	1.680	1.703
60	1.543	1.553	1.564	1.576	1.589	1.603	1.619	1.637	1.657	1.680
70	1.526	1.536	1.547	1.559	1.572	1.587	1.603	1.621	1.641	1.665
80	1.513	1.523	1.534	1.546	1.559	1.574	1.590	1.609	1.629	1.653
90	1.503	1.513	1.524	1.536	1.550	1.564	1.581	1.599	1.620	1.643
100	1.494	1.505	1.516	1.528	1.542	1.557	1.573	1.592	1.612	1.636

Upper Critical Values of the F-Distribution for df_1 Numerator Degrees of Freedom and df_2 Denominator Degrees of Freedom

5% Significance Level

$$F_{0.05}\ (df_1, df_2)$$

df_2 \ df_1	1	2	3	4	5	6	7	8	9	10
1	161.448	199.500	215.707	224.583	230.162	233.986	236.768	238.882	240.543	241.882
2	18.513	19.000	19.164	19.247	19.296	19.330	19.353	19.371	19.385	19.396
3	10.128	9.552	9.277	9.117	9.013	8.941	8.887	8.845	8.812	8.786
4	7.709	6.944	6.591	6.388	6.256	6.163	6.094	6.041	5.999	5.964
5	6.608	5.786	5.409	5.192	5.050	4.950	4.876	4.818	4.772	4.735
6	5.987	5.143	4.757	4.534	4.387	4.284	4.207	4.147	4.099	4.060
7	5.591	4.737	4.347	4.120	3.972	3.866	3.787	3.726	3.677	3.637
8	5.318	4.459	4.066	3.838	3.687	3.581	3.500	3.438	3.388	3.347
9	5.117	4.256	3.863	3.633	3.482	3.374	3.293	3.230	3.179	3.137
10	4.965	4.103	3.708	3.478	3.326	3.217	3.135	3.072	3.020	2.978
11	4.844	3.982	3.587	3.357	3.204	3.095	3.012	2.948	2.896	2.854
12	4.747	3.885	3.490	3.259	3.106	2.996	2.913	2.849	2.796	2.753
13	4.667	3.806	3.411	3.179	3.025	2.915	2.832	2.767	2.714	2.671
14	4.600	3.739	3.344	3.112	2.958	2.848	2.764	2.699	2.646	2.602
15	4.543	3.682	3.287	3.056	2.901	2.790	2.707	2.641	2.588	2.544
16	4.494	3.634	3.239	3.007	2.852	2.741	2.657	2.591	2.538	2.494
17	4.451	3.592	3.197	2.965	2.810	2.699	2.614	2.548	2.494	2.450
18	4.414	3.555	3.160	2.928	2.773	2.661	2.577	2.510	2.456	2.412
19	4.381	3.522	3.127	2.895	2.740	2.628	2.544	2.477	2.423	2.378
20	4.351	3.493	3.098	2.866	2.711	2.599	2.514	2.447	2.393	2.348
21	4.325	3.467	3.072	2.840	2.685	2.573	2.488	2.420	2.366	2.321
22	4.301	3.443	3.049	2.817	2.661	2.549	2.464	2.397	2.342	2.297
23	4.279	3.422	3.028	2.796	2.640	2.528	2.442	2.375	2.320	2.275
24	4.260	3.403	3.009	2.776	2.621	2.508	2.423	2.355	2.300	2.255
25	4.242	3.385	2.991	2.759	2.603	2.490	2.405	2.337	2.282	2.236

26	4.225	3.369	2.975	2.743	2.587	2.474	2.388	2.321	2.265	2.220
27	4.210	3.354	2.960	2.728	2.572	2.459	2.373	2.305	2.250	2.204
28	4.196	3.340	2.947	2.714	2.558	2.445	2.359	2.291	2.236	2.190
29	4.183	3.328	2.934	2.701	2.545	2.432	2.346	2.278	2.223	2.177
30	4.171	3.316	2.922	2.690	2.534	2.421	2.334	2.266	2.211	2.165
31	4.160	3.305	2.911	2.679	2.523	2.409	2.323	2.255	2.199	2.153
32	4.149	3.295	2.901	2.668	2.512	2.399	2.313	2.244	2.189	2.142
33	4.139	3.285	2.892	2.659	2.503	2.389	2.303	2.235	2.179	2.133
34	4.130	3.276	2.883	2.650	2.494	2.380	2.294	2.225	2.170	2.123
35	4.121	3.267	2.874	2.641	2.485	2.372	2.285	2.217	2.161	2.114
36	4.113	3.259	2.866	2.634	2.477	2.364	2.277	2.209	2.153	2.106
37	4.105	3.252	2.859	2.626	2.470	2.356	2.270	2.201	2.145	2.098
38	4.098	3.245	2.852	2.619	2.463	2.349	2.262	2.194	2.138	2.091
39	4.091	3.238	2.845	2.612	2.456	2.342	2.255	2.187	2.131	2.084
40	4.085	3.232	2.839	2.606	2.449	2.336	2.249	2.180	2.124	2.077
41	4.079	3.226	2.833	2.600	2.443	2.330	2.243	2.174	2.118	2.071
42	4.073	3.220	2.827	2.594	2.438	2.324	2.237	2.168	2.112	2.065
43	4.067	3.214	2.822	2.589	2.432	2.318	2.232	2.163	2.106	2.059
44	4.062	3.209	2.816	2.584	2.427	2.313	2.226	2.157	2.101	2.054
45	4.057	3.204	2.812	2.579	2.422	2.308	2.221	2.152	2.096	2.049
46	4.052	3.200	2.807	2.574	2.417	2.304	2.216	2.147	2.091	2.044
47	4.047	3.195	2.802	2.570	2.413	2.299	2.212	2.143	2.086	2.039
48	4.043	3.191	2.798	2.565	2.409	2.295	2.207	2.138	2.082	2.035
49	4.038	3.187	2.794	2.561	2.404	2.290	2.203	2.134	2.077	2.030
50	4.034	3.183	2.790	2.557	2.400	2.286	2.199	2.130	2.073	2.026
60	4.001	3.150	2.758	2.525	2.368	2.254	2.167	2.097	2.040	1.993
70	3.978	3.128	2.736	2.503	2.346	2.231	2.143	2.074	2.017	1.969
80	3.960	3.111	2.719	2.486	2.329	2.214	2.126	2.056	1.999	1.951
90	3.947	3.098	2.706	2.473	2.316	2.201	2.113	2.043	1.986	1.938
100	3.936	3.087	2.696	2.463	2.305	2.191	2.103	2.032	1.975	1.927

Appendix VIII (*continued*)

Upper Critical Values of the F-Distribution for df_1 Numerator Degrees of Freedom and df_2 Denominator Degrees of Freedom

5% Significance Level

$$F_{0.05}\ (df_1, df_2)$$

df_2 \ df_1	11	12	13	14	15	16	17	18	19	20
1	242.983	243.906	244.690	245.364	245.950	246.464	246.918	247.323	247.686	248.013
2	19.405	19.413	19.419	19.424	19.429	19.433	19.437	19.440	19.443	19.446
3	8.763	8.745	8.729	8.715	8.703	8.692	8.683	8.675	8.667	8.660
4	5.936	5.912	5.891	5.873	5.858	5.844	5.832	5.821	5.811	5.803
5	4.704	4.678	4.655	4.636	4.619	4.604	4.590	4.579	4.568	4.558
6	4.027	4.000	3.976	3.956	3.938	3.922	3.908	3.896	3.884	3.874
7	3.603	3.575	3.550	3.529	3.511	3.494	3.480	3.467	3.455	3.445
8	3.313	3.284	3.259	3.237	3.218	3.202	3.187	3.173	3.161	3.150
9	3.102	3.073	3.048	3.025	3.006	2.989	2.974	2.960	2.948	2.936
10	2.943	2.913	2.887	2.865	2.845	2.828	2.812	2.798	2.785	2.774
11	2.818	2.788	2.761	2.739	2.719	2.701	2.685	2.671	2.658	2.646
12	2.717	2.687	2.660	2.637	2.617	2.599	2.583	2.568	2.555	2.544
13	2.635	2.604	2.577	2.554	2.533	2.515	2.499	2.484	2.471	2.459
14	2.565	2.534	2.507	2.484	2.463	2.445	2.428	2.413	2.400	2.388
15	2.507	2.475	2.448	2.424	2.403	2.385	2.368	2.353	2.340	2.328
16	2.456	2.425	2.397	2.373	2.352	2.333	2.317	2.302	2.288	2.276
17	2.413	2.381	2.353	2.329	2.308	2.289	2.272	2.257	2.243	2.230
18	2.374	2.342	2.314	2.290	2.269	2.250	2.233	2.217	2.203	2.191
19	2.340	2.308	2.280	2.256	2.234	2.215	2.198	2.182	2.168	2.155
20	2.310	2.278	2.250	2.225	2.203	2.184	2.167	2.151	2.137	2.124
21	2.283	2.250	2.222	2.197	2.176	2.156	2.139	2.123	2.109	2.096
22	2.259	2.226	2.198	2.173	2.151	2.131	2.114	2.098	2.084	2.071
23	2.236	2.204	2.175	2.150	2.128	2.109	2.091	2.075	2.061	2.048
24	2.216	2.183	2.155	2.130	2.108	2.088	2.070	2.054	2.040	2.027
25	2.198	2.165	2.136	2.111	2.089	2.069	2.051	2.035	2.021	2.007
26	2.181	2.148	2.119	2.094	2.072	2.052	2.034	2.018	2.003	1.990
27	2.166	2.132	2.103	2.078	2.056	2.036	2.018	2.002	1.987	1.974
28	2.151	2.118	2.089	2.064	2.041	2.021	2.003	1.987	1.972	1.959

29	2.138	2.104	2.075	2.050	2.027	2.007	1.989	1.973	1.958	1.945
30	2.126	2.092	2.063	2.037	2.015	1.995	1.976	1.960	1.945	1.932
31	2.114	2.080	2.051	2.026	2.003	1.983	1.965	1.948	1.933	1.920
32	2.103	2.070	2.040	2.015	1.992	1.972	1.953	1.937	1.922	1.908
33	2.093	2.060	2.030	2.004	1.982	1.961	1.943	1.926	1.911	1.898
34	2.084	2.050	2.021	1.995	1.972	1.952	1.933	1.917	1.902	1.888
35	2.075	2.041	2.012	1.986	1.963	1.942	1.924	1.907	1.892	1.878
36	2.067	2.033	2.003	1.977	1.954	1.934	1.915	1.899	1.883	1.870
37	2.059	2.025	1.995	1.969	1.946	1.926	1.907	1.890	1.875	1.861
38	2.051	2.017	1.988	1.962	1.939	1.918	1.899	1.883	1.867	1.853
39	2.044	2.010	1.981	1.954	1.931	1.911	1.892	1.875	1.860	1.846
40	2.038	2.003	1.974	1.948	1.924	1.904	1.885	1.868	1.853	1.839
41	2.031	1.997	1.967	1.941	1.918	1.897	1.879	1.862	1.846	1.832
42	2.025	1.991	1.961	1.935	1.912	1.891	1.872	1.855	1.840	1.826
43	2.020	1.985	1.955	1.929	1.906	1.885	1.866	1.849	1.834	1.820
44	2.014	1.980	1.950	1.924	1.900	1.879	1.861	1.844	1.828	1.814
45	2.009	1.974	1.945	1.918	1.895	1.874	1.855	1.838	1.823	1.808
46	2.004	1.969	1.940	1.913	1.890	1.869	1.850	1.833	1.817	1.803
47	1.999	1.965	1.935	1.908	1.885	1.864	1.845	1.828	1.812	1.798
48	1.995	1.960	1.930	1.904	1.880	1.859	1.840	1.823	1.807	1.793
49	1.990	1.956	1.926	1.899	1.876	1.855	1.836	1.819	1.803	1.789
50	1.986	1.952	1.921	1.895	1.871	1.850	1.831	1.814	1.798	1.784
60	1.952	1.917	1.887	1.860	1.836	1.815	1.796	1.778	1.763	1.748
70	1.928	1.893	1.863	1.836	1.812	1.790	1.771	1.753	1.737	1.722
80	1.910	1.875	1.845	1.817	1.793	1.772	1.752	1.734	1.718	1.703
90	1.897	1.861	1.830	1.803	1.779	1.757	1.737	1.720	1.703	1.688
100	1.886	1.850	1.819	1.792	1.768	1.746	1.726	1.708	1.691	1.676

Appendix VIII (*continued*)

Upper Critical Values of the F-Distribution for df_1 Numerator Degrees of Freedom and df_2 Denominator Degrees of Freedom

1% Significance Level

$$F_{0.01} \ (df_1, df_2)$$

df_2 \ df_1	1	2	3	4	5	6	7	8	9	10
1	4052.19	4999.52	5403.34	5624.62	5763.65	5858.97	5928.33	5981.10	6022.50	6055.85
2	98.502	99.000	99.166	99.249	99.300	99.333	99.356	99.374	99.388	99.399
3	34.116	30.816	29.457	28.710	28.237	27.911	27.672	27.489	27.345	27.229
4	21.198	18.000	16.694	15.977	15.522	15.207	14.976	14.799	14.659	14.546
5	16.258	13.274	12.060	11.392	10.967	10.672	10.456	10.289	10.158	10.051
6	13.745	10.925	9.780	9.148	8.746	8.466	8.260	8.102	7.976	7.874
7	12.246	9.547	8.451	7.847	7.460	7.191	6.993	6.840	6.719	6.620
8	11.259	8.649	7.591	7.006	6.632	6.371	6.178	6.029	5.911	5.814
9	10.561	8.022	6.992	6.422	6.057	5.802	5.613	5.467	5.351	5.257
10	10.044	7.559	6.552	5.994	5.636	5.386	5.200	5.057	4.942	4.849
11	9.646	7.206	6.217	5.668	5.316	5.069	4.886	4.744	4.632	4.539
12	9.330	6.927	5.953	5.412	5.064	4.821	4.640	4.499	4.388	4.296
13	9.074	6.701	5.739	5.205	4.862	4.620	4.441	4.302	4.191	4.100
14	8.862	6.515	5.564	5.035	4.695	4.456	4.278	4.140	4.030	3.939
15	8.683	6.359	5.417	4.893	4.556	4.318	4.142	4.004	3.895	3.805
16	8.531	6.226	5.292	4.773	4.437	4.202	4.026	3.890	3.780	3.691
17	8.400	6.112	5.185	4.669	4.336	4.102	3.927	3.791	3.682	3.593
18	8.285	6.013	5.092	4.579	4.248	4.015	3.841	3.705	3.597	3.508
19	8.185	5.926	5.010	4.500	4.171	3.939	3.765	3.631	3.523	3.434
20	8.096	5.849	4.938	4.431	4.103	3.871	3.699	3.564	3.457	3.368
21	8.017	5.780	4.874	4.369	4.042	3.812	3.640	3.506	3.398	3.310
22	7.945	5.719	4.817	4.313	3.988	3.758	3.587	3.453	3.346	3.258
23	7.881	5.664	4.765	4.264	3.939	3.710	3.539	3.406	3.299	3.211
24	7.823	5.614	4.718	4.218	3.895	3.667	3.496	3.363	3.256	3.168
25	7.770	5.568	4.675	4.177	3.855	3.627	3.457	3.324	3.217	3.129
26	7.721	5.526	4.637	4.140	3.818	3.591	3.421	3.288	3.182	3.094

27	7.677	5.488	4.601	4.106	3.785	3.558	3.388	3.256	3.149	3.062
28	7.636	5.453	4.568	4.074	3.754	3.528	3.358	3.226	3.120	3.032
29	7.598	5.420	4.538	4.045	3.725	3.499	3.330	3.198	3.092	3.005
30	7.562	5.390	4.510	4.018	3.699	3.473	3.305	3.173	3.067	2.979
31	7.530	5.362	4.484	3.993	3.675	3.449	3.281	3.149	3.043	2.955
32	7.499	5.336	4.459	3.969	3.652	3.427	3.258	3.127	3.021	2.934
33	7.471	5.312	4.437	3.948	3.630	3.406	3.238	3.106	3.000	2.913
34	7.444	5.289	4.416	3.927	3.611	3.386	3.218	3.087	2.981	2.894
35	7.419	5.268	4.396	3.908	3.592	3.368	3.200	3.069	2.963	2.876
36	7.396	5.248	4.377	3.890	3.574	3.351	3.183	3.052	2.946	2.859
37	7.373	5.229	4.360	3.873	3.558	3.334	3.167	3.036	2.930	2.843
38	7.353	5.211	4.343	3.858	3.542	3.319	3.152	3.021	2.915	2.828
39	7.333	5.194	4.327	3.843	3.528	3.305	3.137	3.006	2.901	2.814
40	7.314	5.179	4.313	3.828	3.514	3.291	3.124	2.993	2.888	2.801
41	7.296	5.163	4.299	3.815	3.501	3.278	3.111	2.980	2.875	2.788
42	7.280	5.149	4.285	3.802	3.488	3.266	3.099	2.968	2.863	2.776
43	7.264	5.136	4.273	3.790	3.476	3.254	3.087	2.957	2.851	2.764
44	7.248	5.123	4.261	3.778	3.465	3.243	3.076	2.946	2.840	2.754
45	7.234	5.110	4.249	3.767	3.454	3.232	3.066	2.935	2.830	2.743
46	7.220	5.099	4.238	3.757	3.444	3.222	3.056	2.925	2.820	2.733
47	7.207	5.087	4.228	3.747	3.434	3.213	3.046	2.916	2.811	2.724
48	7.194	5.077	4.218	3.737	3.425	3.204	3.037	2.907	2.802	2.715
49	7.182	5.066	4.208	3.728	3.416	3.195	3.028	2.898	2.793	2.706
50	7.171	5.057	4.199	3.720	3.408	3.186	3.020	2.890	2.785	2.698
60	7.077	4.977	4.126	3.649	3.339	3.119	2.953	2.823	2.718	2.632
70	7.011	4.922	4.074	3.600	3.291	3.071	2.906	2.777	2.672	2.585
80	6.963	4.881	4.036	3.563	3.255	3.036	2.871	2.742	2.637	2.551
90	6.925	4.849	4.007	3.535	3.228	3.009	2.845	2.715	2.611	2.524
100	6.895	4.824	3.984	3.513	3.206	2.988	2.823	2.694	2.590	2.503

Appendix VIII (continued)

Upper Critical Values of the F-Distribution for df_1 Numerator Degrees of Freedom and df_2 Denominator Degrees of Freedom

1% Significance Level

$$F_{0.01}\ (df_1, df_2)$$

df_2 \ df_1	11	12	13	14	15	16	17	18	19	20
1	6083.35	6106.35	6125.86	6142.70	6157.28	6170.12	6181.42	6191.52	6200.58	6208.74
2	99.408	99.416	99.422	99.428	99.432	99.437	99.440	99.444	99.447	99.449
3	27.133	27.052	26.983	26.924	26.872	26.827	26.787	26.751	26.719	26.690
4	14.452	14.374	14.307	14.249	14.198	14.154	14.115	14.080	14.048	14.020
5	9.963	9.888	9.825	9.770	9.722	9.680	9.643	9.610	9.580	9.553
6	7.790	7.718	7.657	7.605	7.559	7.519	7.483	7.451	7.422	7.396
7	6.538	6.469	6.410	6.359	6.314	6.275	6.240	6.209	6.181	6.155
8	5.734	5.667	5.609	5.559	5.515	5.477	5.442	5.412	5.384	5.359
9	5.178	5.111	5.055	5.005	4.962	4.924	4.890	4.860	4.833	4.808
10	4.772	4.706	4.650	4.601	4.558	4.520	4.487	4.457	4.430	4.405
11	4.462	4.397	4.342	4.293	4.251	4.213	4.180	4.150	4.123	4.099
12	4.220	4.155	4.100	4.052	4.010	3.972	3.939	3.909	3.883	3.858
13	4.025	3.960	3.905	3.857	3.815	3.778	3.745	3.716	3.689	3.665
14	3.864	3.800	3.745	3.698	3.656	3.619	3.586	3.556	3.529	3.505
15	3.730	3.666	3.612	3.564	3.522	3.485	3.452	3.423	3.396	3.372
16	3.616	3.553	3.498	3.451	3.409	3.372	3.339	3.310	3.283	3.259
17	3.519	3.455	3.401	3.353	3.312	3.275	3.242	3.212	3.186	3.162
18	3.434	3.371	3.316	3.269	3.227	3.190	3.158	3.128	3.101	3.077
19	3.360	3.297	3.242	3.195	3.153	3.116	3.084	3.054	3.027	3.003
20	3.294	3.231	3.177	3.130	3.088	3.051	3.018	2.989	2.962	2.938
21	3.236	3.173	3.119	3.072	3.030	2.993	2.960	2.931	2.904	2.880
22	3.184	3.121	3.067	3.019	2.978	2.941	2.908	2.879	2.852	2.827
23	3.137	3.074	3.020	2.973	2.931	2.894	2.861	2.832	2.805	2.781
24	3.094	3.032	2.977	2.930	2.889	2.852	2.819	2.789	2.762	2.738
25	3.056	2.993	2.939	2.892	2.850	2.813	2.780	2.751	2.724	2.699
26	3.021	2.958	2.904	2.857	2.815	2.778	2.745	2.715	2.688	2.664
27	2.988	2.926	2.871	2.824	2.783	2.746	2.713	2.683	2.656	2.632
28	2.959	2.896	2.842	2.795	2.753	2.716	2.683	2.653	2.626	2.602
29	2.931	2.868	2.814	2.767	2.726	2.689	2.656	2.626	2.599	2.574

30	2.906	2.843	2.789	2.742	2.700	2.663	2.630	2.600	2.573	2.549
31	2.882	2.820	2.765	2.718	2.677	2.640	2.606	2.577	2.550	2.525
32	2.860	2.798	2.744	2.696	2.655	2.618	2.584	2.555	2.527	2.503
33	2.840	2.777	2.723	2.676	2.634	2.597	2.564	2.534	2.507	2.482
34	2.821	2.758	2.704	2.657	2.615	2.578	2.545	2.515	2.488	2.463
35	2.803	2.740	2.686	2.639	2.597	2.560	2.527	2.497	2.470	2.445
36	2.786	2.723	2.669	2.622	2.580	2.543	2.510	2.480	2.453	2.428
37	2.770	2.707	2.653	2.606	2.564	2.527	2.494	2.464	2.437	2.412
38	2.755	2.692	2.638	2.591	2.549	2.512	2.479	2.449	2.421	2.397
39	2.741	2.678	2.624	2.577	2.535	2.498	2.465	2.434	2.407	2.382
40	2.727	2.665	2.611	2.563	2.522	2.484	2.451	2.421	2.394	2.369
41	2.715	2.652	2.598	2.551	2.509	2.472	2.438	2.408	2.381	2.356
42	2.703	2.640	2.586	2.539	2.497	2.460	2.426	2.396	2.369	2.344
43	2.691	2.629	2.575	2.527	2.485	2.448	2.415	2.385	2.357	2.332
44	2.680	2.618	2.564	2.516	2.475	2.437	2.404	2.374	2.346	2.321
45	2.670	2.608	2.553	2.506	2.464	2.427	2.393	2.363	2.336	2.311
46	2.660	2.598	2.544	2.496	2.454	2.417	2.384	2.353	2.326	2.301
47	2.651	2.588	2.534	2.487	2.445	2.408	2.374	2.344	2.316	2.291
48	2.642	2.579	2.525	2.478	2.436	2.399	2.365	2.335	2.307	2.282
49	2.633	2.571	2.517	2.469	2.427	2.390	2.356	2.326	2.299	2.274
50	2.625	2.562	2.508	2.461	2.419	2.382	2.348	2.318	2.290	2.265
60	2.559	2.496	2.442	2.394	2.352	2.315	2.281	2.251	2.223	2.198
70	2.512	2.450	2.395	2.348	2.306	2.268	2.234	2.204	2.176	2.150
80	2.478	2.415	2.361	2.313	2.271	2.233	2.199	2.169	2.141	2.115
90	2.451	2.389	2.334	2.286	2.244	2.206	2.172	2.142	2.114	2.088
100	2.430	2.368	2.313	2.265	2.223	2.185	2.151	2.120	2.092	2.067

Adapted from the National Institute of Standards and Technology Engineering Statistics Handbook.

Appendix IX

Deming's Condensation of the 14 Points for Management

1. Create constancy of purpose toward improvement of product and service, with the aim to become competitive and to stay in business and to provide jobs.
2. Adopt the new philosophy. We are in a new economic age. Western management must awaken to the challenge, must learn its responsibilities, and must take on leadership for change.
3. Cease dependence on inspection to achieve quality. Eliminate the need for inspection on a mass basis by building quality into the product in the first place.
4. End the practice of awarding business on the basis of a price tag. Instead, minimize total cost. Move toward a single supplier for any one item, on a long-term relationship of loyalty and trust.
5. Improve constantly and forever the system of production and service, to improve quality and productivity, and thus to constantly decrease costs.
6. Institute training on the job.
7. Institute leadership. The aim of supervision should be to help people and machines and gadgets to do a better job. Supervision of management is in need of overhaul, as well as supervision of production workers.
8. Drive out fear so that everyone may work effectively for the company.
9. Break down barriers between departments. People in research, design, sales, and production must work as a team to foresee problems of production and in use that may be encountered with the product or service.
10. a. Eliminate slogans, exhortations, and targets for the workforce asking for zero defects and new levels of productivity. Such exhortations only create adversarial relationships, as the bulk of the causes of low

quality and low productivity belong to the system and thus lie beyond the power of the workforce.

  b. Eliminate work standards (quotas) on the factory floor. Substitute leadership.

  c. Eliminate management by objective. Eliminate management by numbers and numerical goals. Substitute leadership.

11. Remove barriers that rob the hourly worker of his or her right to pride of workmanship. The responsibility of supervisors must be changed from sheer numbers to quality.

12. Remove barriers that rob people in management and in engineering of their right to pride of workmanship. This means, inter alia, abolishment of the annual or merit rating and of management by objective.

13. Institute a vigorous program of education and self-improvement.

14. Put everybody in the company to work to accomplish the transformation. The transformation is everybody's job.

http://www.deming.org/theman/teachings02.html

Appendix X

Scorecard for Performance Reporting

Six Sigma Project Scorecard for Performance Reporting

Define	Measure	Analyze	Improve	Control	Percent Complete
Identify Process Owner	Develop Existing Process Flow (Deployment Flowchart)	Validate Project Scope and Process Flow	Identify/Prioritize Root Causes	Develop Process Control Plan	10%
Develop Project Charter & Problem Statement	Implement 5S Housekeeping	Identify Causes of Variation	Screen Potential Causes (DoE)	Implement Control Plan	20%
Develop Stakeholder Analysis	Identify Areas of the 7 Waste Streams	Cause & Effect Diagram	Identify Improvement Actions	Issue Revised Procedures	30%
Develop CTQ Diagram	Develop Data Collection Plan	Box Plot of X and/or Y Variables	Prioritize Improvement Actions	Revalidate Cause and Effect of X's to Achieve Y's	40%
Establish SIPOC Diagram	Measurement Systems Analysis	Multi-Vari Analysis	Develop Implementation Plan	Implement Mistake Proofing	50%
Establish Project Metrics and Baseline Performance	Plot Output (Y) Data over Time (Run Chart/Control Chart)	Histogram of Input (X) Variables	Cost-Benefit Analysis	Monitor Process Performance (Run Chart/Control Chart)	60%
Establish Goals (percent or actuals)	Pareto High Impact Causes/ Process Inputs	Scatter Diagram/Correlation	Risk Assessment/FMEA	Develop before/after Summary Table of Performance/Metrics	70%
Develop Value Stream Map/ Value-Added Analysis	Establish Histogram of Existing Data	Hypothesis Testing	Establish Operating Tolerance for Input Variables	Identify Opportunities to Leverage Improvements	80%
Establish Potential Project Cost Savings/Avoidance	Establish Process Capability of Existing Data	ANOVA	Develop New Process Flow (Deployment Flowchart)	Hand-Off to Process Owner	90%
Review with Sponsor	Review with Sponsor	Review with Sponsor	Review with Sponsor	Review with Sponsor	100%

Overall Project Completion Percentage%:

Sponsor Signature   Sponsor Signature   Sponsor Signature   Sponsor Signature   Sponsor Signature

■ = Completed   ☐ = In Process   ■ = Not Started

Appendix XI

Scorecard for Performance Reporting
(Partly Completed Example)

Six Sigma Project Scorecard for Performance Reporting
(Partly Completed Example)

Define	Measure	Analyze	Improve	Control	Percent Complete
Identify Process Owner	Develop Existing Process Flow (Deployment Flowchart)	Validate Project Scope and Process Flow	Identify/Prioritize Root Causes	Develop Process Control Plan	10%
Develop Project Charter & Problem Statement	Implement 5S Housekeeping	Identify Causes of Variation	Screen Potential Causes (DOE)	Implement Control Plan	20%
Develop Stakeholder Analysis	Identify Areas of the 7 Waste Streams	Cause & Effect Diagram	Identify Improvement Actions	Issue Revised Procedures	30%
Develop CTQ Diagram	Develop Data Collection Plan	Box Plot of X and/or Y Variables	Prioritize Improvement Actions	Revalidate Cause and Effect of X's to Achieve Y's	40%
Establish SIPOC Diagram	Measurement Systems Analysis	Multi-Vari Analysis	Develop Implementation Plan	Implement Mistake Proofing	50%
Establish Project Metrics and Baseline Performance	Plot Output (Y) Data over Time (Run Chart/Control Chart)	Histogram of Input (X) Variables	Cost-Benefit Analysis	Monitor Process Performance (Run Chart/Control Chart)	60%
Establish Goals (percent or actuals)	Pareto High Impact Causes/ Process Inputs	Scatter Diagram/Correlation	Risk Assessment/FMEA	Develop before/after Summary Table of Performance/Metrics	70%
Develop Value Stream Map/ Value-Added Analysis	Establish Histogram of Existing Data	Hypothesis Testing	Establish Operating Tolerance for Input Variables	Identify Opportunities to Leverage Improvements	80%
Establish Potential Project Cost Savings/Avoidance	Establish Process Capability of Existing Data	ANOVA	Develop New Process Flow (Deployment Flowchart)	Hand-Off to Process Owner	90%
Review with Sponsor	Review with Sponsor	Review with Sponsor	Review with Sponsor	Review with Sponsor	100%

Overall Project Completion Percentage%:

Sponsor Signature

Sponsor Signature

Sponsor Signature

Sponsor Signature

Sponsor Signature

= Completed   = In Process   = Not Started

Index